Food, Crop Pests, and the Environment

The need and potential
for biologically intensive
integrated pest management

Edited by

Frank G. Zalom and William E. Fry

D1368465

APS PRESS
The American Phytopathological Society
St. Paul, Minnesota, USA

These reports were prepared in support of the National Integrated Pest Management Forum, Washington, D.C. To make the information available in a timely and economical fashion, this book has been reproduced directly from computer-generated copy submitted in final form to APS Press by the editors of the volume. No editing or proofreading has been done by the Press.

Cover (clockwise from upper left): An apple orchard with weedy ground cover, field corn infected with corn smut, cotton attacked by a boll weevil, and a potato damaged by the Columbia root-knot nematode as examples of cropping systems in which pests are studied in relation to their crop hosts for integrated pest management (photographs by Jack Kelly Clark, courtesy of the University of California Statewide Integrated Pest Management Project).

Library of Congress Catalog Card Number: 92-81857
International Standard Book Number: 0-89054-140-X

Printed in the United States of America on acid-free paper

The American Phytopathological Society
3340 Pilot Knob Road
St. Paul, Minnesota 55121-2097, USA

TABLE OF CONTENTS

FOREWORD

The reports presented in this volume resulted from the thoughtful analysis of scores of practitioners, industry representatives,
growers, environmentalists, scientists, and federal and state agency
personnel. They are acknowledged at the end of each report for their
contributions as members of the "Action Teams" or "workgroups"
for the few cropping systems analyzed. The process began in
summer 1990 when the Environmental Protection Agency (EPA) and
United States Department of Agriculture (USDA) initiated a joint
analysis of the potential for adoption of integrated pest management
(IPM) in United States agriculture. It was decided that individuals
knowledgeable about pest problems in the production of corn/
soybeans, cotton, fruit, or vegetables should describe the current
status of IPM in each of these commodities from a national perspective, and should then provide a "blueprint" for the types of advances
that could be expected with additional research.

Each commodity committee had two co-chairs who shared
responsibility for convening each committee, and for preparing
reports. Committee members provided data, contributed portions of
each report, and critiqued each report. In addition to the committee
members, other individuals were solicited for their reactions, information and criticisms. The reports were revised to reflect these
inputs.

In late summer 1991, the committees met with other representatives of state and federal agencies, industry, environmental groups,
and private practitioners at a conference at the Wingspread Conference Center near Racine, Wisconsin. At that time, predictions
concerning the types of research needed and possible completion
times were developed. These are reflected in the reports.

A major additional function of the commodity committees was
to identify constraints to the implementation of IPM in production
agriculture. These were collated in a report by Dr. Edward Glass,
and provide information for policy-makers to consider. These
constraints are the subject of a forum on IPM sponsored by the EPA
and USDA which will be held in Washington, D.C. in June, 1992.
Commodity committee members felt that constraints relating to
policy and market were among the most serious constraints limiting
the more wide-spread adoption of IPM.

The reports represent the 1992 version of a series of "snapshots"
of the potential for technical improvements achieved via IPM
research to resolve pest-related problems in agriculture. Other
reports have been prepared during the past two decades, and there

are significant similarities among the reports. However, the 1992 versions represent a much stronger emphasis on nonpesticide tactics. These include cultural procedures (some of which until recently were considered to be too unusual or unconventional for adoption), biocontrols, and host resistance. A message that is reinforced throughout these reports is that environmental preservation, food safety, and health of farm workers and farmers, are significantly important factors in modern agriculture.

The authors of these reports acknowledge and thank the many people who contributed significantly to the development of these reports. These include the hundreds of people who responded to questionnaires, participated in meetings, and critiqued drafts of reports. They also include the many personnel in the EPA and the USDA who tirelessly worked with a steering committee and the commodity co-chairs. Without the strong leadership from these agencies, IPM adoption would be further from realization. Richard Parry and Robert Riley of the USDA, and Steven Johnson and Diana Horne of the EPA were particularly instrumental in creating an atmosphere of interagency cooperation which has characterized the effort.

The authors would also like to thank the USDA Cooperative State Research Service (CSRS) for helping to defray the costs of printing this publication, and Joyce Strand of the UC Statewide IPM Project for her efforts in assembling its final draft.

William E. Fry
Department of Plant Pathology
Cornell University
March 6, 1992

Food, Crop Pests,
and the Environment

Chapter 1

INTEGRATED PEST MANAGEMENT: ADDRESSING THE ECONOMIC AND ENVIRONMENTAL ISSUES OF CONTEMPORARY AGRICULTURE

Frank G. Zalom, Director
Statewide Integrated Pest Management Project
University of California
Davis, CA 95616-8621

Richard E. Ford, Professor and Chair
Department of Plant Pathology
University of Illinois
Champaign/Urbana, IL 61801

Raymond E. Frisbie, Director
Center for Biologically Intensive
Integrated Pest Management
Texas A&M University
College Station, TX 77843-2138

C. Richard Edwards, Professor
Department of Entomology
Purdue University
West Lafayette, IN 47907

James P. Tette, Director
New York State IPM Program
NYS Agricultural Experiment Station
Cornell University
Geneva, NY 14456

Pests including disease-causing organisms, insects, mites, nematodes, weeds, and vertebrates lower the quality and yield of agricultural products when left unmanaged. Control of pests and their resulting damage has been an objective of farmers since humans first began cultivating crops and raising livestock. The majority of these activities have been moderately successful. However, evolution of agricultural practices, and evolution of pests create a constantly changing series of new problems that need resolution. Fortunately, the practice of pest control has continued to advance making it possible to produce a variety of crops in different regions of the world with fairly predictable results.

Agricultural practices, including those involving pest control, have had significant impacts on the environment. Clearing and burning forests and native grasslands for new fields, clean cultivation to prevent competition from weeds and residual inoculum of insects and diseases, and application of agricultural chemicals to enhance quality and yield each have had measurable impacts on the environment. Advances in agriculture in general and in pest control technologies in particular have allowed the world's population to increase to its present level. The tragedy of famine today is perhaps less a question of worldwide food production in agriculture than of politics, economic development, and inadequate transportation and distribution of food resources. Certainly in many areas of the world, people have never enjoyed greater diversity or abundance of food and fiber.

Prior to the late 1940's American farmers relied upon non-chemical pest control methods such as crop rotation, tillage, and hand-removal of pests. Pesticidal materials which were available prior to the introduction of synthetic pesticides contained elements such as copper, sulfur, lead, antimony and arsenic, and compounds of botanical origin such as nicotine and pyrethrum. Many of these materials were toxic to humans and were expensive to produce in quantity, therefore the availability of many, particularly the botanicals, was limited. Equipment for their handling and application was relatively unsophisticated or lacking. Overall pesticide use was low relative to contemporary levels. During this period, an awareness of the effectiveness of biological controls was developing. This awareness was particularly enhanced following the successful 1890 introduction by the United States Department of Agriculture (USDA) of the Vedalia beetle from Australia to California to control an introduced pest of citrus, the cottony cushion scale (11).

PESTICIDE IMPACTS

The development of synthetic chemicals during World War II, in combination with improvements in application technology, dramatically increased the potential for controlling pests. Many pest control researchers focused on applying and perfecting the technology. Pesticide development, production and use became institutionalized. Farmers became increasingly dependent upon pesticides because they were reliable, economical and saved labor. In certain crops and regions, pesticides have even been used for purposes of expanding soil conservation. For example, in 1985, 95% of corn and soybean acreage was treated with herbicides compared to only 40% in 1970 (20). Much of this increase is due to the widespread use of no-till practices in the midwestern United States which helped to address issues of soil erosion. In many situations, pesticides permitted the production of crops in areas where they could not be previously grown, during times of the year when pest pressure is greatest, and without the need for strategies of rotation or fallow periods. Extended seasons and postharvest pesticide use permitted fresh products to be available to consumers for periods far longer than ever before, and reduced the risk of potentially dangerous toxins developing in processed products.

Although considerable attention has been focused on proposed legislation that would affect the availability of certain pesticides or crop uses of pesticides, other factors have been responsible for limiting their availability or usefulness to growers. These factors include pest biology, effect on nontarget species and human health issues. A description of some of these factors follows.

Secondary Outbreaks

Many plant-feeding insects do not significantly damage agricultural crops because they are kept under natural control by other organisms. However, some of these organisms which exert natural control can be killed by chemical applications, resulting in the plant-feeding species becoming a new pest species which is then often referred to as a secondary pest.

Resistance

Pest resistance to a chemical can develop rapidly because the life cycles of many pests are relatively short. In a pest population, some individuals will be genetically resistant to pesticides applied for their

3

control. Even when a high percentage of the population is killed, these few individuals survive, reproduce and pass the genes which have allowed them to resist the pesticide to the succeeding generation. Thus, a pest population develops which can be managed only with higher dosages of the chemical; finally the pesticide will no longer control the pest. Most pesticides have a limited effective life. Resistance has been reported in almost 500 species of insects and mites, 100 species of plant pathogens, 50 species of weeds, 5 species of rodents, and 2 species of nematodes (6).

Effect Upon Non-Target Organisms

Pesticides generally kill a broad spectrum of plant-associated organisms, only a fraction of which are the target pests. Some of these effects impact agriculture directly. For example, pesticide-induced losses of commercial honey bees in the United States total between $20 and $50 million annually. Various organisms which provide a food source for biocontrol agents when the pests are not abundant, and those organisms which function in decomposition and mineral recycling in the soil may also be affected. In addition, under some conditions pesticides have phytotoxic effects on agricultural crops resulting in reduced yields or quality.

Organisms which are peripherally associated with agricultural systems have been known to be affected as well. Persistent soluble pesticides may filter into the soil and be carried away by irrigation or rain water into streams where they enter the food chain via consumption by microorganisms, invertebrates, fish and possibly higher organisms. The best known example is DDT which was banned in the United States in 1972 when it was implicated in mortality of wild bird populations by causing thin-shelled eggs.

Human Exposure

People may come in contact with pesticides most commonly during application and entry into treated areas. Such exposure may be mitigated when proper protective clothing is worn and equipment is used during the mixing and application of pesticides. Pesticide label guidelines regarding reentry of treated fields and harvest intervals are intended to protect agricultural workers and consumers, but workers do not always follow those guidelines, creating some risk.

Other Environmental Hazards

The impact of pesticide use on the environment is complex, and not always fully understood. Some pesticides are extremely slow to break down due to natural processes, and lose their toxicity very slowly. A few pesticides have been found to move through soil, and have been detected in groundwater. As the ability to detect minute amounts of chemicals has increased, the potential extent of contamination has come under increasing scrutiny. Unfortunately, the ability to detect contamination has surpassed our ability to accurately assess possible impacts.

REGULATION

Concern for the impact of pesticides on human health and the environment has resulted in increased regulatory action by both federal and state legislatures and regulatory agencies. Some of the laws that have or will directly affect the availability of pesticides include the United States Endangered Species Act and the Federal Insecticide, Fungicide and Rodenticide Act as amended in 1988 (FIFRA 88). State laws also affect pesticide use and availability. California, for example, has enacted several laws targeting pesticides alone or in combination with other chemicals including the Birth Defects Prevention Act, the Safe Drinking Water and Toxic Enforcement Act, the Pesticide Contamination Prevention Act, and the Toxic Air Contaminants Act. In addition, pesticide manufacturers have voluntarily withdrawn several products in anticipation of additional regulatory requirements. Also, new toxicological problems are occasionally identified in older pesticide compounds, which result in their withdrawal from the market.

Perhaps the greatest potential impact on pesticides has resulted from FIFRA 88. It requires the accelerated reregistration of pesticides for which the United States Environmental Protection Agency (EPA) did not have complete registration data. More than 4000 pesticide uses on food crops alone are subject to reregistration under this law (18). Federal reregistration is scheduled for completion in 1997, but some pesticides will lose registrations before 1997 if their registrants do not agree to provide data or do not pay reregistration fees. Under FIFRA 88, cancellations will occur if the EPA determines a pesticide should not be registered for a use, or when registrant companies voluntarily withdraw pesticide registrations.

Increasing pest problems which continue to influence pesticide efficacy and increasing public concern and regulatory complexity make the future of synthetic organic pesticides uncertain. It may be

5

assumed that individual registrations will continue to be withdrawn or canceled, and proposed laws and regulations will result in further cancellations. Pesticide development will continue, but new registrations are becoming increasingly more difficult and expensive to obtain. Clearly, a need exists to assess the impact of reduced availability of pesticides in our various cropping systems, and to identify environmentally sound pest control approaches.

INTEGRATED PEST MANAGEMENT

As early as the 1950's, pest researchers wrote of the danger of relying on a single pest control technology such as pesticides. These individuals recognized the potential contribution of pesticides, but also viewed the technology in the broader context of the agricultural production system. The philosophy of "integrated control" or "integrated pest management" was developed.

Integrated Pest Management (IPM) is an ecologically-based pest control strategy which is part of the overall crop production system. "Integrated" because all appropriate methods from multiple scientific disciplines are combined into a systematic approach for optimizing pest control. IPM tactics employed must be compatible with each other and with social, environmental and economic factors. "Integrated" inherently implies interdisciplinary efforts which are crucial for the successful development and application of IPM programs. "Pests" include all organisms which impact the production of food and fiber, and IPM strategies may be developed which are appropriate for pests of human health, forests and urban areas as well. "Management" implies acceptance of pests as inevitable components, at some population level of an agricultural system.

The IPM approach is to use a series of tactics to reduce overall pest populations. Pesticides or other control tactics are applied only after all other relevant tactics have been deployed or when their need is justified by knowledge of pest biology, established decision guidelines, and the results of field monitoring. Ideally, IPM programs consider all available management options including taking no action.

The concept of the treatment threshold is a key element of IPM systems. It has been repeatedly referred to as the "economic threshold" meaning the pest population density at which control measures must be applied to prevent an increasing pest population from reaching the point where crop loss would exceed the cost of control. However, the economics of pest control can be quite ephemeral and although the term "economic threshold" has often been criticized, the concept is quite valid. Damage would be tolerated by producers

6

until the threshold is reached. Determining a control action threshold requires knowledge of pest biology and crop physiology as they relate to the environment, naturally occurring biological controls and the effects of possible control actions. They may vary from region to region, crop cultivar to crop cultivar, and even between fields depending on specific agricultural practices.

IPM TACTICS

Although IPM employs a variety of pest control tactics, biological approaches are often stressed. In fact, the first mention of "integrated control" in the scientific literature involved the selective use of pesticides in walnuts to preserve parasites of the walnut aphid (13). However, integrated strategies have been part of some pest management systems without a label for the past 100 years. For example, suppression of "potato decline" caused by viruses was managed by growing seed tubers in isolation and also suppressing insect vectors of the viruses. Cultural controls are also important IPM tactics, and include a broad range of production practices intended to render the crop and related environments less favorable for the pest. Proper tillage practices, crop rotation and water management are effective cultural controls in the management of many types of pests including soil pathogens, nematodes, weeds, vertebrates and soil arthropods. The destruction of crop residues is important in the management of many pests. Chemical controls can be used as IPM tactics as well. When they are available, selective pesticides which kill only the target species are generally desirable because they are the least disruptive to the crop ecosystem. The choice of pesticide should also consider the safety of workers, and the potential for pest resistance developing when only one chemical or class of chemical is used extensively.

IPM computer modeling efforts have been undertaken for several crop systems and pests, and have served as a framework for integrating information from various disciplines. Some models which predict aspects of plant growth and pest population levels have become important in managing pests in the field, and are widely used.

INSTITUTIONALIZING IPM

The process of developing and implementing IPM programs has occurred at some level for almost 40 years. The term IPM was coined and IPM research was conducted by some scientists in the 1950's, a few of whom were pioneers in bringing pest management knowledge and information directly to growers as private pest control advisers.

IPM research continued in the 1960's, without a formal structure for funding or coordination of this work. The pest management consulting profession was relatively small during the 1960's which reflected the widespread application of pesticides to agricultural crops on a routine "calendar" basis. Increased regulation on the use of pesticides, problems associated with pesticide efficacy and increasing costs of pesticides and their application increased the need for IPM research and educational efforts, and for trained pest management professionals who could apply IPM strategies and tactics directly in farmers fields.

IPM was first accepted institutionally in the early 1970's when individuals in government, universities, and various agricultural industry groups called for its increased development and application. During this time, several significant reports were completed. A historical perspective of the roles and responsibilities of federal and state agencies in pest management was presented as a preface to a detailed study on IPM by the Office of Technology Assessment. Eight workgroups and two large advisory committees contributed to the report. In 1972, the Council on Environmental Quality (CEQ) published *Integrated Pest Management* (3), and defined IPM as:

> "...an approach that employs a combination of techniques to control the wide variety of potential pests that may threaten crops. It involves maximum reliance on natural pest population controls, along with a combination of techniques that may contribute to suppression—cultural methods, pest-specific diseases, resistant crop cultivars, sterile insects, attractants, augmentation of parasites or predators, or chemical pesticides as needed."

Following publication of this report, President Richard Nixon directed CEQ to provide initial leadership in overall pest management policy, citing in his Environmental Message the urgent need to regulate pesticide use due to the perception of harmful persistence in the environment and food chain magnification. The CEQ envisioned the need to integrate many of the IPM components already in place. It described the available techniques and embraced the concept of IPM and it committed the federal government to practice IPM principles and to fund its research and implementation.

Later in the decade, President Jimmy Carter again called for a commitment to IPM, defining it in his 1979 Environmental Message as follows:

"IPM uses a system approach to reduce pest damage to tolerable levels through a variety of techniques, including natural predators and parasites, genetically resistant hosts, environmental modifications and, when necessary and appropriate, chemical pesticides. IPM strategies generally rely first upon biological defenses against pests before chemically altering the environment."

The National Science Foundation (NSF) served as lead agency for funding of the Huffaker Project, a coordinated program of research on IPM for several key agricultural systems and human health. The Huffaker Project was succeeded by a more broadly based Consortium for Integrated Pest Management (CIPM) project which involved 15 universities researching the IPM of four crops: cotton, soybeans, apples and alfalfa. As funding ended for CIPM in the mid 1980's, the Experiment Station Committee on Policy (ESCOP) Directors at Land Grant Universities established a system of regional IPM research projects.

In the early 1970's the USDA Extension Service served as a lead agency for a series of IPM Pilot Projects to enhance the delivery of IPM technologies. These early efforts concentrated upon a few major cropping systems of the United States. Subsequent efforts have included a variety of urban and agricultural applications.

Cotton has been a focus of major IPM research and extension activities in all previous national IPM initiatives. It is therefore significant to note that insecticide use on cotton has decreased since the extension IPM Pilot Projects were initiated in 1971. In 1971, an average of 5.8 pounds per acre were applied to cotton. This decreased to 5.5 pounds per acre in 1976, and 1.5 pounds per acre in 1982 (4). Of total pounds of insecticides applied to major field crops in 1976, cotton accounted for 49%. By 1982, cotton accounted for only 24%. However, at least some of this reduction might be attributed to the use of new pesticides used at lower rates such as the pyrethroids, which were introduced during this time. In Texas alone, 82% of the cotton growers studied as part of the National Survey of Extension IPM Programs, coordinated by the Virginia Cooperative Extension Service, were identified as IPM users (17).

The Texas IPM program in cotton has a long history dating back to the original extension IPM Pilot Projects for cotton and tobacco which was initiated in 1971 (7). In a pilot project, state extension services hired and trained scouts who provided IPM services to growers subscribing to the programs. Today, field scouts who are part of the Texas Pest Management Association, a farmer operated nonprofit association, provide IPM monitoring services to 2,500

Texas cotton growers farming 958,000 acres. California also participated in this pilot project and by 1973 over 200,000 acres of cotton, or 22% of the state total, were serviced by private pest management consultants (15). About 80% of the California cotton acreage at present. The primary role of scouts hired as part of the early extension IPM programs was to monitor pest abundance or incidence, and then to report their findings to their Cooperative Extension Service (CES) supervisors who would work with growers to determine the best IPM tactic for use. The system of IPM scouting did not persist in large part because CES funding withered at this time when IPM programs would have required a doubling or tripling of personnel to do this labor-intensive job. Fortunately, consultants, many trained by CES specialists, have taken up some of these activities.

Dramatic declines in pesticide use have been observed in other crops targeted in the extension Pilot Projects. Peanuts have received considerable research and extension attention throughout the southern United States. In 1971, 87% of peanut acreage was treated with an average of 3.9 pounds of insecticides per acre. By 1982, only 48% of peanut acreage was treated, and the average pounds per acre was less than 0.8 (4). Similarly, corn has been an important target for CES efforts. In Indiana approximately 90% of the continuous corn acreage (1.4 million acres) is treated with a soil insecticide for corn rootworm control. By rotating corn with another crop, producers can virtually eliminate the need for a rootworm insecticide. Acreage in continuous corn has been reduced from 40% to 27% of all corn acreage over the past 15 years. This reduction can be attributed to: 1) nearly equivalent net return from producing corn or soybeans, 2) numerous agronomic advantages associated with a rotational system, 3) changes in government programs, and 4) promotion of corn/soybean rotation by Purdue University's IPM Specialists. This amounts to yearly savings in insecticide costs for rootworm control of approximately $7.7 million, and a reduction of about 700,000 pounds of insecticide applied to Indiana soils.

Perennial cropping systems have also benefited from the successful implementation of IPM research and extension programs. California almond production is one example of a perennial crop where a complex of IPM techniques have been widely adopted by growers. Most of the techniques were developed over the decade of the 1970's, and implemented in the early 1980's resulting in reduced crop damage, increased production, and a reduction in total pesticide use of 31.2% (12).

As demonstrated by these case studies among many others, there is strong empirical evidence that when adopted, IPM results in economic benefit to growers and society, typically accompanied by reduction in pesticide use. In spite of this, IPM techniques are far from being adopted universally, even in many cases where convincing empirical evidence exists that IPM would be beneficial. Federal funding for IPM research and extension activities has declined substantially in real terms during the decade of the 1980's. This has been replaced by state support in a few cases, notably in California, Texas, and New York, where special funding from state legislatures has helped create IPM programs which include both competitive grants and permanent IPM extension staff.

We now know that the public and their governmental representatives had greater expectations of rapid replacement of conventional pesticides through the use of IPM than was reasonable to expect. In this regard it is somewhat ironic that the legislative mandate which initiated the original joint research and development effort through the USDA, EPA and NSF ended with the final summary sentence which is as relevant today as then: "The Federal Government can help, but the long-term success of IPM depends upon the states, the universities, the private IPM industry, and ultimately the farmer."

THE NEED FOR IPM HAS NEVER BEEN GREATER

A national survey of farmers conducted in 1990 indicated that closely following their desire for marketing information was their desire for information on weed control, disease control, cultivar selection (many cultivars incorporate pest resistance to reduce losses), insect control and nematode control. All of these were mentioned ahead of their need for information about fertilization, planting, harvesting, water management, land preparation and crop storage. When asked what could help to reduce their production costs the farmers suggested, immediately after breeding for improved yields, that genetic resistance to pests, development of new pesticides/herbicides and biological control agents would be more helpful than even improvements in equipment. Critical in increasing yields in priority order were weed control, breeding and pest control/management ahead of seven other factors. When asked what traits would affect profits, they listed more yield and more disease resistance ahead of eight other factors.

A recently completed quadrennial strategic plan for the State Agricultural Experiment Stations assigned "Safe and Effective Management of Pests" in the 1990's, fourth priority among 31 identified. This issue includes pest-resistant genes, resistance screening, control strategies, IPM, biological control agents, ecological impacts and biological shelf life as key researchable components of pest management. IPM has achieved that high priority status

11

from its history of outstanding research and extension efforts, and from recognition by both scientists and producers that the next quantum leap in plant health and higher yields will happen largely by removing these biological constraints to full yield potentials. The ranking also recognizes that IPM is a realistic approach to managing pests while addressing society's environmental concerns. It is essential then that IPM efforts be supported at a national level, and that renewed emphasis be placed on the development and implementation of IPM by government, universities, the agricultural industry and producers.

Chapter 2

INTEGRATED PEST MANAGEMENT IN THE CORN/SOYBEAN AGROECOSYSTEM

C. Richard Edwards, Professor
Department of Entomology
Purdue University
West Lafayette, IN 47907

Richard E. Ford, Professor and Chair
Department of Plant Pathology
University of Illinois
Champaign/Urbana, IL 61801

INTRODUCTION AND BACKGROUND

The United States corn/soybean agroecosystem is one of the world's largest most intensive farming systems. Nearly 50% of the 325 million acres of cropped land in the United States lies within the Corn Belt. The 10-state Corn Belt, characterized by near optimum environment and resources, produces over 83% of the Nation's corn on 80% of the acreage, and nearly 77% of its soybeans on 73% of the acreage. Soybean, "a new crop" important in the United States during the last 50 years now constitutes more than half of total world production, and is a major export crop. The corn/soybean agroecosystem is primarily a rotation of these two crops, but monocropping of corn is practiced in areas heavily committed to livestock and where the climate restricts soybean harvest to a short period each fall. Wheat and grain sorghum are important rotational crops in some states, with double cropping of wheat followed by soybean in the same year in some southern portions of the Corn Belt states. Availability of management options varies widely with farm size, which ranges from 200 to more than 100,000 acres. Sufficient flexibility must be built into integrated pest management (IPM) efforts to accommodate these wide range of variables. Farms are highly mechanized and efficient, and the cropping system must be considered when developing IPM programs for corn. The major soybean agroecosystem in the southeastern United States involves soybean/corn/forage, soybean/corn/cotton/small grain, soybean/grassland, soybean/rice, and soybean/sugarcane. There is great diversity on many farms including tobacco, peanuts, and vegetable crops, plus hedgerows, forests, and swamps providing numerous wild hosts of soybean pests.

Pest control tactics are designed to disrupt the favorable combination of biotic and environmental factors necessary for pest development. Pest management is an essential part of the crop production system. Pest management strategies emphasize prevention whenever possible because many pests cannot be controlled as effectively if they become established during the cropping season. A combination of tactics is available to reduce the variety of pest threats in the Corn Belt. Current tactics are based principally on chemical and cultural (mechanical) controls for weeds and insects and on cultural controls and genetically resistant plants for diseases.

Based on the above, the key focus of all deliberations was to identify constraints on research, development, delivery and use of IPM in the corn/soybean agroecosystem. Specific constraints are listed in each section. Some constraints common to all sections need not be repeated; some of them will likely be common also to the other commodities. The major common constraints are:

- Lack of adequate funds for staffing, research, training, education, and technology transfer (primarily long-term interdisciplinary efforts);

- The relatively low cash value of corn and soybeans does not allow an adequate profit margin for growers to invest in long-term solutions;

- Current federal farm programs which require farmers to grow specific crops to protect their program base often discourage the use of good pest management practices, such as crop rotation;

- Lack of fundamental research on pests, pest interactions, and cropping systems;

- The stringency of Environmental Protection Agency (EPA) regulations and requirements for testing and registration of non-conventional biologicals causes reluctance in some companies to invest the requisite large sums of money to develop effective IPM technologies;

- While effective IPM is management intensive, biologically enhanced IPM requires even a higher level of management and knowledge; and

- Lack of an education program at the advanced level to supply people with multidisciplinary training.

Pests

IPM for corn targets its efforts on the reduction of about two dozen each of weeds, insects, and diseases. However, only a few of these pests consistently reduce corn yields and are, therefore, of major concern. Herbicides for management of grassy weeds account for most of the pesticide dollars spent by growers. Soil-borne plant diseases are the least manageable with current technologies. Major insect pests are the corn rootworm, the European corn borer and the black cutworm. Corn rootworms cause damage primarily where corn follows corn in the crop rotation.

The economic impact on soybeans is generally ordered by rank with weeds the highest followed by nematodes, diseases, and insects. Because it is difficult to separate the effects of any one pest, the soybean producer must consider the total effect of all pests. The costs of weed management practices, which include herbicides and tillage, are high. Weeds compete directly with the crop for nutrients, water and light. They also interfere with the operation of equipment and harbor insects, plant pathogens and nematodes. Insects may reduce yields by attacking roots, stems, foliage, and/or seed pods. Insect pests of greatest economic importance on soybeans are the complexes that feed on foliage and pods in August and September. Direct costs of insect pests are related primarily to crop losses and expenditures for insecticide application. Nematodes in the soil feed on roots and usually are not noticed until populations increase to damaging levels. Host resistance is now available to help manage the soybean cyst nematode. Soil fumigants are not cost effective and are seldom used. During the last decade the advent and development of some systemic fungicides have aided in the management of some fungal diseases. This has been especially important in assuring the sale of clean seed for planting.

Cultural Pest Management

Cultural practices are an integral part of pest management and are most effective when combined with other measures. Cultural practices used in the Corn Belt to reduce survival, germination, development, or spread of pests include the use of clean, disease-free seed, adjusted planting or harvesting dates, tillage, drainage, crop sequence, crop rotation, plant nutrition (fertilization), and sanitation.

Biological Control

The regulation of pests by natural enemies is one reason many pests seldom reach their full biotic potential in the Corn Belt. Indigenous parasitic or antagonistic biological control organisms control many soil-borne pests of corn and soybeans. Some of these biological control agents can be manipulated by specific cultural practices. Such manipulation by habitat management has been generally as effective as the introduction and establishment of exotic organisms. In some areas, growers have learned the benefits of natural insect enemies, and benefited from not applying insecticides until or if economic thresholds are reached.

Plant Genetic Resistance

Host plant genetic resistance is a primary pest management tactic effective against most diseases. Plant breeders have an impressive record in the development of disease resistant germplasm. An early concern was raised (1) that 66% of the germplasm parentage in use for corn could be traced to 5 inbred lines. This has now broadened to at least 12 inbred lines. Twenty cultivars of soybeans constituted 92% of the acreage planted in the United States, but more importantly, the parentage of all cultivars traces to 10 ancestral types. The National Academy of Sciences report concluded, however, that "lack of genetic diversity probably has not caused special pest problems in corn or soybeans, nor has it led to unusual applications of pesticides." However, the concentration of a large percentage of harvested acres in the Corn Belt in only corn and soybeans, has resulted in the persistence and perhaps increase in some insect, weed and disease problems.

Pest-resistant plants provide a natural, economical, environmentally safe, self-generating system that is generally compatible with

16

other control tactics, is readily accepted by farmers, and has been a primary control tactic for several decades. Pest-resistant cultivars are vital to the control of certain nematodes and plant diseases. The effectiveness of resistance genes accounts for a significantly low usage of pesticides for disease management. Unfortunately no genetic resistance has been identified for a few diseases, such as Septoria leaf blight on soybean. Some diseases such as stalk rots of corn can not be controlled, but are managed only with difficulty by a complex of multiple genes. Resistance to certain insects has been identified and resistant lines are in various stages of development. Certain cultivars with different growth patterns may compete better with weeds. Herbicide tolerance varies among existing soybean cultivars and corn hybrids.

Current IPM Practices

IPM practices are increasingly used in the Corn Belt as producers gain a greater awareness of the advantages of pest control through an integrated approach. Many crop consultants established businesses in the last decade to assist growers in regular monitoring of crop health. The interest in IPM reflects both the recognition of IPM successes and a growing concern for sustainable agricultural production. Recent innovations in pest monitoring greatly improves the efficiency of chemical and cultural control tactics. In this way, IPM can play an important role in minimizing nonpoint source pollution by pesticides. Steadily IPM is being accepted and used by farmers in the Corn Belt. Thus, the farm management system must include effective IPM. Current adoption and use of IPM is limited largely by the lack of basic research information on pest biology, timely biological and weather data, and a data base for accurate pest detection, prediction of pest density, and the relationship of pest density to crop loss. Without these, the prophylactic use of pesticides as "insurance" against pest problems will continue. IPM integrates pest control into crop protection/crop production systems that will reduce the severity of pests, the frequency of pest problems, and pest resistance. Reduction, not elimination of, pesticide use is one of the objectives of IPM, through use of tactics such as pest scouting, resistant host plants, crop rotations, cultural practices, and biological control.

Agricultural Chemicals

Chemical pesticides are important in the management of each complex of corn and soybean pests. Weed management depends more on chemicals than does either insect or nematode management. Pesticides are applied to both soil and seed to provide effective, dependable, and sometimes the only control for some pests or pest complexes. The largest quantities of pesticides used on corn in the Corn Belt are applied to the soil, pre- or post-emergence, for weed and insect control. Plant disease management has depended generally less on chemicals, but the use of foliar-applied and seed-treatment fungicides can increase yields and thus are now more widely used. Nearly all acreage cropped to corn and soybeans receives some sort of chemical application to manage one or more pests. Basically, management of all of the major weeds depends on chemical herbicides that perform best when used in addition to good cultural practices, rather than as the sole means of control. A shift to reduced tillage for erosion control, moisture retention, and labor and energy efficiency has increased the need for and reliance on some pesticides. Zero-tillage systems may also require fungicides, rodenticides, and higher dosages of insecticides and herbicides because contemporary pesticides may not be as effective in reducing increased perennial weed pressure or insect populations in these systems. Accumulation of crop residue also enhances rodent populations within fields.

New Concerns

Each innovative adjustment to the farming schedule for economic considerations must be monitored to detect potential negative side effects to the agroecosystem. Some of these effects can be anticipated while others appear as surprises.

Exotic Pests
We must maintain exclusion as a high priority strategy to prevent the introduction of pests such as soybean rust, maize streak virus, the soybean pod-borer, and numerous weeds. We have learned from a history of introductions that some insects, weeds and pathogens in a new environment in the absence of natural enemies can lead quickly to the development of epidemics. The European corn borer is a classical example.

Pest Resistance
The evolution of pest resistance to pesticides is a natural phenomenon in response to environmental pressure. Certain races of the soybean cyst nematode and the root-knot nematode now cause serious losses on previously resistant cultivars. Serious levels of resistance have occurred in soybean looper to methomyl and some

plant pathogens have already developed resistance to benomyl as its use increases. Triazine resistant weeds have been identified since the 1970's, and new cases of herbicide resistance are appearing rapidly, especially for new, low-rate herbicides. Strategies for gene deployment and the use of pesticides must be developed carefully to increase the usefulness or "shelf life" of these various tactics.

STATUS, PRACTICES AND CONSTRAINTS

Crop Rotations/Tillage Systems

Crop rotation and tillage systems affect every aspect of crop production. Weed, disease, insect, and fertility management are the most important reasons growers rotate crops or plow and disk the soil. As rotation and tillage intensity increase, the number and diversity of pests generally decrease. Adoption of crop rotation is strongly correlated with farm size. Crop rotation and other alternatives to continuous cropping are most common on small farms. Continuous cropping is most common on the largest farms. Adoption of conservation tillage tends to be inversely proportional to farm size. Conservation tillage is practiced most commonly on large farms and least commonly on small farms. Adoption of conservation tillage is influenced by farmer participation in government programs.

Crop rotation provides other economic and environmental benefits to corn/soybean producers. Crops grown in rotation are generally more productive than crops grown continuously due to improved soil moisture, soil tilth, nutrient availability and pest control. The benefits of particular tillage systems are more complex to evaluate. Conventional plow-disk systems aid in pest management in general, but unfortunately promote soil erosion and surface runoff on sloping lands. Minimum and no-till systems greatly reduce soil erosion and surface runoff, but require increased pest management inputs. The type, duration and management of crop rotations are influenced by regional and local environmental, edaphic, agronomic and economic conditions.

Life cycles of insects that attack plant roots are particularly susceptible to disruption by removal of the food source when crops are rotated. The strategy of altering planting date to avoid injury during peak pest population pressures could be equally as effective as rotations in some instances. Rootworms are managed by rotation of corn with other crops, because the pest is limited in mobility and corn is its only food source during the larval stage. However, some northern corn rootworms have adjusted to crop rotations by having their diapause extend through a growing season. Additionally, crop rotations may increase populations of other insect larvae. Crop rotation is important for control of soybean cyst nematodes, but it seems to have relatively less impact on insect pests of soybeans.

The ability of pathogens to overwinter, to disseminate inoculum, and to infect alternate crops determines the effect of rotation on disease incidence. Where soilborne pathogens are involved, crop rotation has little influence on disease incidence.

Weeds are less effectively controlled by crop rotation alone than are other pests. Crop rotation patterns affect weed populations mainly through the effectiveness of herbicides in specific crops. However, when growers shift from one tillage system to another, weed populations change. Adoption of conservation tillage has been associated with increased annual and perennial weed problems, including the appearance in no-tillage of woody perennials not commonly associated with crop production. As tillage intensity decreases, the spectrum of weed species changes.

Tillage destroys unwanted vegetation, disrupts soil insect habitats, and buries plant residues carrying inoculant of disease organisms. Conventional tillage permits mechanical cultivation of weeds. Eliminating tillage reduces the disruption of weeds, insects, and disease organisms, requiring increased reliance on other management options. Vertebrate pests and slugs may increase in no-tillage systems. These pests are encountered primarily in regions where nonagricultural habitats are intermixed with cultivated fields.

Tillage systems are currently determined mainly by soil conservation, fertility and weed control requirements. As tillage intensity decreases the amount of crop residue left on the soil surface increases. This residue helps control erosion and reduces surface runoff, but favors the build-up of pests and makes weeds more difficult to control. As a result, pest control requirements tend to increase in reduced tillage systems. Growers who want to conserve soil and reduce pollution of surface water, while reducing pesticide inputs, face a major dilemma in managing weeds and to a lesser extent other pests. On erodible land, the 30% surface residue requirement will be difficult to obtain in soybeans except in no-tillage or in rotation with high residue crops.

Available Now
- Crop rotations may reduce incidence of stalk rots indirectly because leaf blights make corn more susceptible to stalk rots. Eyespot, anthracnose, and southern corn leaf blight are

generally less common on corn grown in rotation than on corn grown continuously.

- Crop rotation decreases damage to soybeans from soybean cyst nematodes, *Alternaria* leafspot, anthracnose, bacterial blight, *Cercospora* leafspot, pod and stem blight, mildews, *Septoria* brown spot, stem canker and other diseases. Many bacterial and fungal pathogens that survive on crop debris eventually die if residues decay before the next crop of soybeans is planted.

- Population increases of many weed species are interrupted by rotating to more competitive crops or to crops in which the weeds are more easily controlled.

- Soybean cyst nematode is effectively managed by rotation to non-host crops.

- Burial of crop residue is the most effective method to reduce damage caused by slugs and vertebrate pests.

- Plow-tillage and surface-tillage control armyworm through early season destruction of plants attractive for oviposition.

- Reducing early season weed growth minimizes cutworm problems in all tillage systems.

- Stalk borer damage is reduced by tillage, or in conservation tillage, by late summer weed control (especially giant ragweed and grasses) to eliminate oviposition sites.

- Seedcorn maggot damage is reduced by tillage or by elimination of surface mulches.

- Lesser cornstalk borer damage is avoided by use of no-tillage systems, maintaining high amounts of surface mulch, especially from double cropping.

Current Research
- On ridge tillage, for areas where suitable.

- Combination of banded pesticides with limited tillage as a means of reducing pesticide use.

21

- On the "rotation effect" related to nitrogen availability to grain crops following legumes.

- Comparison of suitability of various rotations for local needs.

- Applications of standard IPM concepts to weeds.

- Weed seed banks/weed population studies as related to rotations.

Constraints to Further Development

- Lack of funding for pest ecology and biology research with application to field crop production, especially for long-term crop rotations; major agricultural research funding (USDA Competitive Grants; USDA Low Input Sustainable Agriculture (LISA) Grants Program, etc.) do not address the critical needs in this area.

- University promotion and tenure policies discourage faculty from engaging in crop rotation research.

- Crop rotation research is management oriented rather than technology oriented.

- Lack of fundamental research on interactions among pests, especially on weed biology.

- Lack of information on germplasm potential especially with respect to cover crops.

- Decreased understanding and appreciation of agricultural research by an increasingly urban electorate.

- The desire for a "quick fix" with cheaper and easier to use short-term pesticides to solve pest problems, when required modifications of tillage and crop rotations are long-term with high initial cost.

- Traditional emphasis is strong throughout agriculture on pest "control" as opposed to pest "management" that may involve rotations.

- Growers plan based on how the farm program affects short-term economics, rather than on long-range crop rotations directed toward pest management.

- Many factors complicate crop rotation decisions including tillage, herbicide carryover, conservation needs, market projections, input costs, risk of pest damage, government programs and crop enterprise needs. Integrating this information to arrive at the optimum crop rotation pattern is difficult. However, the prime consideration of growers who might wish to rotate crops is an economic one based upon anticipated crop prices and expected net returns. Thus, short-term financial consequences are more important determinants of crop rotations than are agronomic or ecological ones.

- Lack of sound research-based information about crop rotations as opposed to information based on past experience or tradition.

Biological Control

Biological control is the use of living organisms, or products thereof, to reduce the magnitude and extent of problems caused by pests. Although it is widely believed that successful biological control is difficult to achieve in annual row crops, especially in regions where little permanent or natural vegetation can be found, many successful examples exist in both corn and soybeans, mostly for insects. Enthusiasm must be tempered, however, by the knowledge that no practical biological control has yet been found for many pests. Weeds are by far the most troublesome class of pests, and relatively few biologicals are available for use against them.

Examples of classical biological control that are available or show strong promise include importation and release (with permanent establishment) of predators, parasites and pathogens. A related successful approach involves mass rearing and release of the natural enemies on an as-needed basis. Although predators and parasites require specialized release techniques, microorganisms can usually be applied in a manner similar to that used for chemical pesticides. Viruses, fungi, and especially the endotoxin of the bacterium *Bacillus thuringiensis*, commonly called *Bt*, are being used this way. *B. thuringiensis* is highly commercialized, whereas most others are still under research and development.

23

Recently, some new and innovative biological approaches have been developed. For example, certain insect pests require and use specialized weed habitats during part of their life cycle. By removing or modifying these weeds from the vicinity of the corn or soybean field, the pest's life cycle can be disrupted greatly reducing crop damage. Most recently scientists have discovered that the *Bt* gene for endotoxin production can be introduced into plants, thus effectively building a corn or soybean plant that produces its own internal biological control agent. Examples of this approach appear in the sub-section on *Genetic Engineering*.

Kairomones, chemicals produced by one species and which communicate beneficially to another species, hold promise in attracting natural populations of predators or parasites to fields where their services are desired. Additionally, plant kairomones can be used to draw phytophagous insects to poison baits. Allomones, which are chemicals produced by one species and which communicate to another species to the producer's benefit, hold promise for suppressing germination of weed seeds or repelling insect pests. In the simplest case, crop breeding programs would be used to incorporate allomone production into commercial hybrids or cultivars.

Available Now

- *B. thuringiensis* is used as a crop protectant in corn against European corn borer and in soybeans against green cloverworm and soybean looper.

- Naturally occurring insect predators of mites, aphids and caterpillars in both corn and soybeans can be conserved by selective cultural practices. These predators include *Amblyseius, Aphidoletes, Chrysopa,* and *Hippodamia*.

- Selective weed management is effective to suppress insect pests. Examples include eliminating winter annuals from fields where corn will be planted to remove oviposition sites for black cutworm, and mowing or eliminating grass from field borders in the spring to remove armyworm oviposition sites and in the summer to remove European corn borer "action sites."

- Insect parasites and predators are mass-reared for release on an annual basis against some pests. Proven parasites include *Pediobius foveolatus* against Mexican bean beetle and *Trichogramma spp.* against European corn borer. The green lacewing, *Chrysosperla carnea,* is effective against aphids, caterpillar eggs, and mites.

- A fungus *Colletotrichum gloeosporioides* that can be applied as a spray for controlling northern jointvetch, a weed in rice and an occasional pest in soybeans in limited areas.

- Biological control of the musk thistle with the weevil *Rhinocyllus conicus.*

Current Research

The research areas presented below are expected to produce results that can be put into practice within 5 years:

- Incorporating the gene for *Bt* toxin into endophytes living in corn plants for protection against European corn borer, corn earworm, armyworm and corn rootworm.

- Formulations of *B. thuringiensis* which include attractants and feeding stimulants.

- Use of *B. thuringiensis* as a crop protectant against corn rootworm and against bean leaf beetle and Mexican bean beetle in soybeans.

- Corn rootworm adult control using attractants and arrestants to deliver an insecticide precisely to the target, thereby greatly reducing the amount of pesticides used per acre.

- Managing disease-inhibiting soils which contain microbial antagonists that suppress growth of root pathogens.

The following areas of current research are expected to yield implementable results primarily beyond 1996:

- Improved biological control of selected weed species using insects or plant pathogens (fungi, nematodes, bacteria and/or viruses).

- Corn rootworm control by the nematode *Steinernema carpocapsae.*

- Endophytic *Beauveria bassiana* in corn for control of European corn borer.

25

- Control of soil-borne plant pathogens by strategic use of animal manures.

- Use of living mulches.

Constraints

- Tendency for growers to seek immediate solutions which leads them to select fast-acting chemicals instead of slower-acting biologicals.

- The relatively low profit margin of corn and soybean crops do not allow growers to invest heavily in long-term solutions.

- Funding agencies and administrators often expect scientists and research programs to show positive results in 1 to 3 years, whereas research on biological systems often takes 5 to 10 years before conclusions can be drawn.

- Factual data on pest levels and the damage they cause are scant because few comprehensive surveys are conducted. This makes it difficult to prioritize research and development programs or to justify changes in the status quo.

- Field research on mobile species (especially vertebrates, many insect pests, and most insect predators and parasites) cannot be done on small plots. For some species the minimum useful plot size exceeds one acre.

- Shortage of scientists dedicated to and properly trained for researching biologically intensive methods, especially in weed science.

- Many of the more serious pests in corn and soybeans live most of their life cycle below the soil surface as insect eggs and larvae, weed seeds and roots, or fungal spores. Below ground research is difficult and expensive.

- Biologically intensive IPM cannot be packaged easily and sold by the private sector because many aspects involve decision making that require management time but do not involve purchasing an input.

- Inability to patent beneficial organisms, other than genetically altered ones.

- Regulatory policies requiring registration restrict interstate movement and release of biological pesticides.

- Evolutionary changes among pest species continue, thus whatever management is implemented will become less effective with time. This requires a continual development of new tactics and new strategies.

Plant Genetic Resistance

Plant breeding, particularly the selection for genetic resistance to various pests, has been one key to production increases for both corn and soybeans. Plant breeders develop cultivars for the conditions of specific production zones and incorporate available pest tolerance or resistance into commercial lines. Genetic resistance is the preferred method of reducing damage from pathogens, nematodes, and insects. Unfortunately, the resistance of crop plants to weeds is much more complex, involving the interaction of many crop traits such as growth habit, rate of vegetative growth, canopy structure and many other factors; the genetic basis of these traits is not well understood.

Pest resistant cultivars may come with additional costs to the grower, especially in initial releases. Frequently, these costs are in the form of yield reductions. For example, where the soybean cyst nematode is present, susceptible soybean cultivars suffer yield losses of 50% or more while resistant cultivars produce a reasonable yield. The first resistant soybean cultivars typically yielded 10% less than susceptible ones in the absence of the soybean cyst nematode. Newer soybean cyst nematode resistant cultivars now yield as well as the susceptible ones in the absence of the soybean cyst nematode. Producers must often make difficult management decisions about when to use a resistant cultivar and which resistance genes are necessary. These decisions must be based on the best information available on the pest before the growing season starts. Frequently, growers purchase seed of a pest resistant cultivar as insurance, much as certain pesticides are used prophylactically.

Cultivars with pest resistance provide a pest control alternative which avoids possible health and environmental hazards associated with the use of many chemicals. Of course, care must be taken to insure that the health aspects of the cultivar as food are not affected by the pest resistance mechanism. Resistance or tolerance is often the only economical option available for reducing the damage caused by certain pests. Examples of such pests include soil-borne organisms, stalk borers, viruses and nematodes. However, sources of breeding material resistant to pests are not always available. Screening lines for resistance is expensive, time consuming and no genetic tolerance is known for numerous pests. Also, pest populations, especially certain pathogens, are notorious for rapidly overcoming resistance, particularly if selection pressure is heavy. The corn/soybean cropping system using cultivars with single gene resistance is particularly prone to changes in the pest population. *The average useful life span for cultivars resistant to corn borer (not single) and some leaf blights is less than 10 years. It takes at least that long to develop a replacement cultivar.*

Classical plant breeding for genetic resistance to pests remains the best tool available for enhancing the successful application of IPM systems to row crops. Certain plants with resistance to phytophagous insects produce allochemicals that are detrimental to parasitoids. Generally, host plant resistance against pests fits well into any management practice and does not require specialized farming equipment. It is also relatively inexpensive and predictable, usually not a hazard to people or the environment, simple to apply and available to all growers. Biotechnology will enhance the use of genetic resistance to pests by providing for improved selection of systems, information on gene function, ability to modify resistance genes and new sources of genetic material.

Available Now
- Limited resistance to leaf feeding and tolerance to sheath collar feeding by European corn borer is available in some corn hybrids and continues to improve.

- Corn and soybean cultivars with specific resistance to one or more of the following: soybean cyst nematode, fungi, viruses, bacteria and insects.

- Corn and soybean cultivars that emerge or mature outside the window of pest activity.

- Cultivars that contain genes for generalized pest resistance or tolerance, rapid root growth, strong stalks, hairy pods, and similar host plant resistance factors.

- Cultivars which tolerate moisture, temperature or salinity stress and therefore are less susceptible to pest invasion.

Current Research
- Corn and soybean breeding represents the core research and development effort of the largest seed companies in the United States.

- Private and public programs focus on the identification of new lines containing pest resistance.

- Computer technology and biotechnology have added tools to develop more efficient procedures to identify resistant lines.

- Genetic engineering research offers potential new sources of genetic material.

- Researchers are investigating the genetic responses of weed competition for light especially in the seedling stage as a prelude to development of competitive cultivars of crops.

- Improved methods of storage and preservation of genetic material.

- Introduction and genetic improvement of new alternative crops.

Constraints
- Lack of availability of long-term funding.

- Lack of support for breeder education programs.

- Loss of genetic diversity.

- Profitability of private sector research and development.

Genetic Engineering

Biotechnology is the use of aspects of biology, including genetic engineering, to develop useful products and services. Genetic engineering is the introduction or modification of genes in an organism using recombinant DNA technology. Products of genetic engineering include transgenic plants, microorganisms, and animals as well as compounds derived from these organisms.

The technology embodied in genetic engineering will impact agriculture significantly. Corn will benefit significantly from the application of this technology since hybridization technology makes it necessary for producers to buy seed rather than save it. Hybridization has enabled seed companies to develop exclusive cultivars that are not easily duplicated by competing companies. Thus, a competitive, profitable market opportunity drives leading edge research and development in seed corn companies. Genetic engineering provides the technology to refine further, but not make obsolete, the commercial seed market for corn and could also impact the marketing of soybean seed.

Genetic engineers can address plant/pest interactions at the molecular level. This understanding will enable scientists to devise measures more specific and less harmful to nontarget organisms. These measures for pest management could include new synthetic chemicals, as well as transformed plants marketed as proprietary seed and improved biological pesticides.

Available Now
- Genes that confer resistance to pests including viruses and insects.

- Tools and methods for studying pathogenicity at the molecular level.

- Methods for transforming corn and soybeans.

- Methodology for transforming bacteria to enhance biopesticide properties.

- Restriction fragment length polymorphisms (RFLP) and random amplified polymorphic DNA (RAPD) gene-mapping techniques for characterizing both microbial and crop genomes, particularly corn, and for more rapid selection for specific genetic characteristics.

- Diagnostic tests for some plant diseases, especially viruses.

- Recombinant *Bt* products are nearing commercialization for control of European corn borer.

Current Research
- Application of molecular biology to the study of plant/pest interactions.

- Incorporation of pest resistance genes into commercial corn lines.

- Isolation and characterization of genes that confer pest resistance.

- Further characterization and mapping of genomes of corn and soybean.

- Development of more effective methods for the creation of transgenic plants and microbial pesticide organisms.

- Improved gene expression systems.

Constraints
- Regulations on research at an early stage hamper product development.

- The regulatory pathway for commercialization is undefined partly because of delays or uncertainty in the federal rule making process which results in increased state oversight.

- Restrictions on interstate movement of microorganisms.

- International regulatory diversity and uncertainty.

- Lack of highly trained regulatory staff.

- High cost of development.

- Poor public acceptance of products of biotechnology.

Decision Making and Implementation

Implementation of biologically intensive IPM will depend heavily on the collection, analyses, integration, and utilization of information. Therefore, sampling and decision-making may involve relatively simple problem identification, integration of historical and real-time data, or complex predictions of future events and management decisions.

Available Now
- Tools for rapid problem identification include traditional manuals, leaflets, color slides, some expert systems software

31

and biotechnical tools (e.g., rapid chemical and immuno-assays).

- Scouting programs (i.e., in situ counts, traps, etc.) and economic thresholds for several corn and soybean insects (e.g., corn rootworm, European corn borer, black cutworm, stalk borer, green cloverworm, and bean leaf beetle), and to a lesser extent for a limited number of diseases caused by fungi and nematodes.

- SOYBEAN/ds©, which is an expert system to aid in diagnosis of a subset of the major soybean diseases.

- Methods to estimate abundance (population densities) of potential vertebrate pests.

- Weed management programs for soybeans designed to optimize the use of postemergence herbicides based on weed scouting and economic thresholds that have been developed primarily in the southeastern United States.

Current Research

- Weed management programs for corn, based largely on sampling weed seeds and seedlings, originally developed in Colorado, and currently being adapted to the North Central Region through a regional research project.

- Development of expert systems for corn and soybean IPM built around corn and soybean growth models.

- Various studies that should lead to improved decision guidelines (i.e., action thresholds) and/or better sampling methods.

Constraints

- Limited resources will necessitate identifying and eliminating barriers to intra- and interdisciplinary research and extension programs.

- Lack of applied research at the systems level.

- Limited applied research on timing of biologically-based management strategies (in comparison with that on "chemically intensive" strategies).

- A higher level of management will be required.

- Lack of adequately trained personnel to collect and process information needed for decision-making.

Pest Resistance Management

The development of resistance in pests to pesticides has caused concern, since the discovery in the 1950's that resistance had developed in several insect species with continued use of chlorinated hydrocarbons. Today resistance in insects has been noted for every major class of insecticide. The same phenomenon was observed in fungal pathogens, e.g., the development of resistance in the apple scab fungal pathogen to dodine, in the peanut leafspot pathogen to benomyl, and in numerous phytopathogenic organisms to the benzimidazole fungicides. The situation became critical when the downy mildew and the potato leaf blight organisms developed resistance to metalaxyl and related acylalanine-type materials. This resulted in the establishment of an unprecedented industry task force to attempt both to minimize and to manage the development of resistance to fungicides. Weeds have developed resistance to the triazine herbicides and more recently to the sulfonylurea and other herbicides. These examples demonstrate clearly the need for immediate, decisive action concerning the entire resistance problem.

Available Now
The following are means of minimizing the development of resistance which apply both to chemical and non-chemical management tools:

- Rotate crops to reduce availability of substrates for pests and pathogens.

- Alternate chemical and non-chemical control methods to slow down or reduce the possibility for development of pest resistance.

- Use cultural practices of cultivation and sanitation.

- Use tolerant or resistant cultivars.

- Interplant crops to provide a diverse gene pool.

The following are methods of minimizing the development of resistance to chemicals and to microbe based biocontrols:

- Alternate the use of chemicals which possess different modes of action.

- Use chemicals which are not site-specific in their mode of action.

- Use chemicals only when the economic threshold is reached.

- Minimize the use of chemicals when the pest population has reached excessive levels.

- Use chemicals on trap crops, in traps, or with a pheromone monitoring program.

- Use chemicals at a minimum effective rate. Do not increase the rate in an effort to broaden control spectrum.

The following methods may minimize the development of resistance to genetically-modified organisms:

- Do not use genetically-induced resistance as the sole means of pest management; use in conjunction with other pest management methods such as chemicals or non-related biologicals.

- Promote germplasm preservation in situ, growers fields and natural areas.

- Intermix genetically-altered seed with non-altered seed.

- Interplant genetically-modified crops with non-modified crops.

- Select for, or insert, multiple genes for resistance.

- Rotate between genetically-modified crops and non-modified crops.

Constraints

- Insufficient awareness/utilization of current means of managing pest resistance.

- Lack of concentrated interdisciplinary research and support of resistance management activities.

- Need for more education and training at all levels.

- Need for more experts in all areas related to resistance management.

- Need for better cooperation and communications among/ between companies, universities, government and non-profit institutions.

- Insufficient focus by the various interest groups on the resistance problem.

- Lack of understanding and support by industry, government, universities and non-profit institutions regarding the severity of the resistance problem.

- Lack of a central function to focus on and coordinate the pest resistance problem.

- Insufficient funds to support the training, education, research, and interdisciplinary effort required for successful resistance management.

- Development of resistance by the soybean cyst nematode to the resistant soybean cultivars.

- Burdensome oversight of technological alternatives and subsequent public apprehension in acceptance of new alternatives.

IPM Technology Transfer and Extension

Cooperative Extension Service (CES) personnel should focus on a higher level of education about biological control and should take an increasing leadership role in the development and delivery of the technology for biologically intensive IPM. As the program evolves, CES should facilitate the coordination of growers, industry, consultants, government agencies and researchers to plan, develop and deliver biologically intensive IPM systems mutually adaptable and usable in today's agriculture.

A movement toward more biologically intensive IPM will require producers to rely more heavily on biological and mechanical controls, host resistance, and cultural management as the primary strategies for pest management. Pesticides are integral in the pest population management system. It is imperative that pesticides be environmentally safe and used only when other strategies are inappropriate. Under pest "outbreak" situations for most pests of corn and soybeans, synthetic chemicals will remain a key line of defense. However, biologically intensive strategies must be developed and used to reduce the chances and/or impact of "outbreaks."

The transition of CES programs in corn and soybeans from chemically-based IPM to more biologically-intensive systems will require a strong commitment by federal and state governments through both extension and research. Biologically intensive IPM programs developed for use by corn and soybean producers must be available to all producers. CES has built a strong foundation for the change to more biologically intensive IPM through its successful track record for IPM program development and implementation.

Available Now

- CES pest management specialists in each state for most pest disciplines.

- State-of-the-art IPM training being conducted by most state CES's.

- Established infrastructure in CES allows for the further development of IPM programs.

- "How to" CES IPM manuals and other program enhancement materials available, although not uniform across states.

- CES newsletters containing information, including pest identification, scouting techniques, pest prediction and forecasting, measuring thresholds, management alternatives, etc., all pertinent components of pest management.

- CES diagnostic centers for pest and damage identification and management information.

- Some regional CES IPM specialists, although inadequate in number, to demonstrate IPM techniques and technologies, deal with the establishment of local pest management programs and react to local pest problems.

- Generally, good interaction of CES specialists within disciplines in each state. Usually good interaction of specialists among pest disciplines.

- A computerized interactive database of all available extension IPM materials is available through CES.

Constraints

- Inadequate funding for enhanced development and transfer of IPM programs.

- Lack of CES IPM personnel to develop and transfer local-level technology which is highly site-specific. Since IPM requires increased management, many more practitioners are needed.

- Lack of individuals in the private sector to provide pest management services to growers. In the future, producers will seek more highly trained technical expertise and a higher level of service.

- Minimal support for adaptive research from university administration for the development of IPM strategies.

- IPM curricula do not provide adequate practical pest management training for use on a regional basis.

- Lack of field oriented IPM research being done at universities.

- Limited amount of research and/or extension effort to develop certain aspects of IPM such as use of biological controls against insect pests and plant pathogens, application of multiple pest thresholds for all pests, application of weed economic thresholds, application of alternative strategies for weed management, use of detection systems for plant pathogens, application of resistant lines for nematode management and research to assess the impact of vertebrates and to develop non-chemical management strategies.

- Lack of ability for CES to show tangible increases in producer profits through IPM program participation since many benefits of IPM are long-term.

- Lack of personnel to do site-specific research and demonstrations.

- Unwillingness on the part of granting agencies to fund long-term research of 5 to 10 years.

- Lack of a CES program supporting private independent practitioners.

Agricultural Chemicals

The development and use of agricultural chemicals represents one of the four major revolutions in agriculture along with mechanization, breeding/genetics and biotechnology. The discovery and adoption of agricultural chemicals for pest management have allowed the growers to 1) better realize the crop yield potential, 2) manage pest outbreaks, 3) reduce the need for on-farm labor, 4) increase the quantity and quality of food, feed and fiber, 5) increase profits, and 6) reduce or maintain the price of food/land. Therefore, agricultural pesticides continue to significantly impact agriculture in terms of quantity, quality and cost of food and feed.

Before the discovery of pesticides, most pests were managed either manually or by rotation of crops. Many crops suffered large yield losses, and some simply could not be grown. Pesticides have allowed the pests that compete with crops to be managed economically and efficiently. Pesticides have reduced crop losses, increased the quality of food and helped maintain a more predictable supply.

Early pesticides were developed with little regard for safety and environmental consequences. A public outcry demanded food and environmental protection. More emphasis was placed on the use of less toxic compounds as more was learned about risks associated with pesticide use and as analytical techniques advanced. Many unsafe pesticides have been removed from the market. Growers can now control essentially all economic pests of corn and soybean in the United States.

Biologically enhanced IPM will not likely eliminate chemical pesticide agents in the control of weeds and diseases. Although not all pathogens which cause plant disease are transmitted by insect vectors, IPM programs may effectively reduce incidence of some diseases through insect control. Changing cultural patterns such as to minimum-, or no-tillage, will change the weed or disease complex, but will not keep others from developing. Instead of trying to eliminate pesticides, IPM approaches should integrate pesticide usage. In addition, management of some weeds by biological organisms will be difficult because weed competition with the crop could possibly be yield-reducing by the time a biological achieved control.

With the advent of more effective and safer compounds, pesticides are likely to continue to be important tools for corn and soybean production even in more biologically intensive IPM programs.

Available Now
A wide variety of chemical, and a few biological, pesticides are presently available. With these available pesticides, along with proper tillage, rotations, etc., most economic pests which attack or compete with corn or soybeans can be adequately managed.

Current Research
- Development of herbicides which can be used post-emergence, thus allowing applications to be better timed, targeted and more effective.

- Evaluation and implementation of reduced rates for some pesticides.

- Discovery and development of biological pesticides.

- More effective timing and placement of the pesticides being used.

- Research continues on chemicals with different modes of action and lower environmental impact.

Constraints
- High cost of developing new pesticides.

- The unpredictable nature of legislation which impacts policy.

- Time and difficulty of obtaining a new registration.

- Social concerns related to the use of pesticides.

- Low economic return on new products due to the many available and effective pesticides on the market. Because of this, smaller companies are not able to afford the high development costs of many of the biological pesticides.

Regulations

Regulation of Pesticides

Pesticides are regulated by the EPA under the Federal Insecticide, Fungicide and Rodenticide Act (FIFRA). Before a pesticide can be marketed, it must be tested by the applicant and the results must be evaluated carefully by EPA scientists. A pesticide is approved or "registered" only after the EPA determines that the use of the pesticide will not have unacceptable risk of adverse effects on humans, domestic animals, wildlife and the environment. Pesticides registered earlier are subject to the same testing and review standards in a process called reregistration.

Many states have passed legislation requiring companies to do health and safety reviews before products can be used. These additional reviews generally focus on uses and environmental factors considered as unique to the region.

Pesticides are used to control pests which reduce crop quantity and quality. Pests must be managed so that an adequate supply of safe and wholesome food and feed can be produced. Under current economic and regulatory conditions and with the current state of scientific knowledge, pesticides effectively aid in the management of most corn and soybean pests. Research must continue on alternative methods of pest control. Alternative methods must be made available to growers who can select the most desirable options. It would not be productive to halt research on agricultural chemicals, or on alternatives, simply because adequate means to control these pests are presently available. We must continue to strive for means to control pests which are effective, economical and safe. Some of the "newer" pesticides are applied at low rates, some as low as a few grams of active ingredient per acre, with much improved environmental and human safety. Lower rates have many benefits, however, use-rate as well as the potential toxicity of the chemical in question are both important when considering safety.

Regulation of Production and Cultural Practices

Federal policy does not impose quotas to control the production of most crops. Instead, support payments have been in place to ensure a minimum payment to growers. This minimum payment acts as a guarantee to growers and encourages them to produce the highest yields possible. This policy has been effective in providing for abundant, high quality, wholesome and inexpensive agricultural products, but is perceived as forcing growers to rely on high levels of outside inputs to produce these high yields. In addition, growers are required to grow certain crops to "maintain" their basis for support payments. Critics claim that these production practices are not sustainable and must be changed.

Adoption of IPM Practices

Yields. A direct correlation generally exists between pest management and increased quantity and quality. For this reason, growers commonly attempt to achieve high or economic levels of pest management in their crops. At some point, continued attempts at control are more expensive than the resulting increased yields.

Weed Control. Weeds are considered an economic pest in all fields of corn and soybeans. In most fields, weeds are the only pests present at economic levels; therefore, growers attempt to manage these pests.

Weeds compete with the crop for nutrients, moisture and light, interfere with mechanical harvesting, and serve as alternate hosts for some insects and diseases. It may be necessary, therefore, to control weeds for reasons other than to increase yields. Growers commonly attempt to manage the pests that progress beyond the economic threshold. Lacking data on economic thresholds for most crop pests, growers attempt to control rather than manage pests.

IPM Systems. Current IPM systems are generally effective only for a single (or possibly a few) pest(s) present in a crop. No single system has been developed that will manage all of the pests in a crop. Few weed management options are available other than herbicides and mechanical cultivation.

Biological controls generally manage, but do not eliminate pests. High levels of pest management are not always maintained, thus yields from fields where alternate or biological pest control measures have been practiced are often lower than from fields where chemical practices have been employed. Yield reduction, however, does not always result in an economic loss to the producer, as input costs may also have been reduced.

IPM Adoption. Growers strive for the maximum crop yields attainable for many reasons. Although some reasons are related to laws and government programs, the bottom line for growers is profit. It may be difficult to persuade growers to adopt alternate pest control methods if such methods result in lower yields. If a grower was to be presented with a production system that could

41

provide equivalent economic returns, most likely it would be widely tested and adopted.

Constraints on Regulations

- Federal and state regulations, production programs and crop grading standards and requirements.

- EPA regulations and requirements for the testing and commercialization of biological pesticides so stringent that they inhibit small companies and public sector scientists from developing them.

- Regulations associated with biotechnology, especially recombinant DNA, appear to be a major unknown which have some effect on the development and use of products that should support biologically intensive IPM.

- Increasingly burdensome costs and time frames associated with the registration of new chemicals prohibits the development of very selective compounds that are desirable for use in IPM systems.

- Restrictions on the interstate movement of biological control material.

- Variability in review policies and procedures at international, federal and state levels.

FUTURE IPM BLUEPRINT

Future Research and Implementation Needs of Biologically Based Alternatives

The corn/soybean production system is possibly the most finely-tuned of any agroecosystem in the United States. Current research from both public and private institutions has contributed to the overall IPM effort which has slowly gained acceptance and momentum in this agroecosystem. Funding is a necessary incentive to enhance the adaptation of research results to practical pest management programs at the grower level. Unanimous consensus ranks lack of funding as a very important constraint on the development and implementation of biologically enhanced IPM.

Product quality has high priority in all aspects of marketing of corn and soybeans as both commodities are used directly in human foods and in animal feeds. Successful IPM programs must consistently produce clean grain, free from contamination by toxic by-products from insects, weeds, and/or pathogens.

Unpredictable weather patterns create unique stresses on both the biological system and on managers in an agroecosystem which relies on natural rainfall. As a result, IPM programs need a multidimensional flexibility which is not required by many irrigated crops. For example, cool, wet springs require measures to offset reduced seed germination, stand vigor and root health, all of which result in slower crop growth and lower yields. Conversely, a wet fall, which impedes or delays harvest, can result in mycotoxin contamination. This creates a completely different problem of low grain quality even in times of record yields.

Implications of Removing Pesticides

Elimination of pesticides would impact primarily weed control in production systems of both corn and soybeans. The major agrichemical market is herbicides. Herbicide use increased in parallel to the strong emphasis on adaptation of ecological farming principles such as reduced tillage to eliminate soil erosion. In reduced tillage systems, soil-borne disease, rodent, and insect complexes have shifted markedly in composition due to the cooler, more moist soils resulting from crop residues allowed to remain on the soil surface. Where once fall plow-down covered debris so that beneficial soil microorganisms could degrade organic matter, it also decreased the viability of insect eggs, fungal and bacterial pathogens. Now more of these pathogens and insects are able to overwinter in soil surface plant debris, emerging in the spring to infect and infest the crop.

The elimination of pesticides would increase the cost of corn and soybean production. Here are examples to illustrate the point.

- Elimination of all herbicides could require up to triple the current number of mechanical cultivations to manage weeds below the economic threshold. This translates to a reduction in cost for herbicides of $20 per acre for corn and $25-30 per acre for soybeans with an increase in cost for each mechanical cultivation required of $6.25 per acre. New machinery requirements and access to more labor creates some unknown costs in this conversion. "Successful nonchemical weed control requires highly refined management skills and is as much an art as a science" (8). It would also increase the loss of priceless topsoil through soil erosion, with a concomitant increase in pollution of surface waters.

- Seedcorn maggots, which may increase in no-till systems, are controlled by the treatment of seed with insecticides. The elimination of the $1 per acre seed treatment could result in a 12% reduction in stand where seedcorn maggots are present which would represent a loss of about 12 bushels per acre.

- Many seedling diseases are controlled by chemical seed treatment, currently a very inexpensive target specific approach to improving stand vigor. Seed treatment is more important for corn than for soybeans as the soybean plant has a built-in capacity to compensate for stand reductions of up to 20% before noticeable yield losses occur. The only known remedy is to plant in warmer soils, i.e., delayed planting to early June translates to reduced yields of approximately 10% for soybeans and 25% for corn in Minnesota (8). Considerable public pressure is placed on farmers to utilize environmentally sensitive practices to protect soil, air and water. Legislating adherence to these practices presents a dilemma the public would frown on if the no-till or minimum till systems were abandoned in favor of top soil eroding mold-board plowing and intensive deep-soil cultivation. Yet, if we eliminate pesticides, which the public also considers desirable, costs of machinery and labor for mechanical cultivation will be added to the long-term permanent loss of environmental resources such as top soil and soil fertility. We cannot place a value on such important resources.

The judicious use of chemicals is currently inextricably integrated into our system for reasons of efficiency, economy and environmental preservation. It is also incumbent on us to assure the public that the chemicals used are environmentally safe. Thus, a sophisticated system of regulatory review and registration is required. Although costly, it is important to maintain a system that engages public confidence.

Regional differences in soil types, temperatures, rainfall and winds dictate the need for adaptability of IPM systems to meet the needs for optimizing productivity of the corn/soybean agroecosystem. These differences also mandate regionally based research and development programs for technology adaptation.

Future research is a fundamental part of improving the corn/soybean agroecosystem and moving it toward a more environmentally secure, sustainable system. In the paragraphs that follow, available and future technology that could contribute to this goal are summarized.

Crop Rotations/Tillage Systems

While little new technology will be available, adaptation of current technology to optimize the interaction of soil saving tillage systems and crop rotations with evolving pest management systems will require considerable new research. Some examples of research needed to answer questions we see arising in the future are:

- Increase use of legumes and other cover crops in rotations.

- Develop an understanding of weed seed dormancy and the responses to biological and mechanical manipulation of seeds selected from the weed seed bank for use in predicting weed emergence under various environmental conditions.

- Develop better models to clarify how tillage and crop rotation influence occurrence of pest resistance, a result of certain IPM practices.

- Develop overseeded or interseeded rotational cover crops both to provide a habitat for beneficial insects and to have allelopathic properties to aid in weed control.

- Study soil compaction and residue cover to be managed more effectively using automated precision tillage and controlled traffic patterns.

- Adopt precision farming, using biologically interfaced sensors coupled with electronic controls to improve the efficiency of delivering precisely timed quantities of inputs that address real-time crop and soil needs.

- Create biologically based simulation models for use to predict pest population dynamics in various crop rotations and tillage combinations as an aid in pest management decision making.

45

Biological Control

Research in this area is increasing rapidly, partly because biological control has been identified as a priority area for USDA Competitive Grants. Some examples of research required to answer questions which have arisen or will arise are:

- Develop proper management of allelopathic crops and cover crops to suppress weeds.

- Do more intensive research on the fundamental biology and ecology of weeds to enhance the long-term management of weeds.

- Understand better responses to pest injury (including insect and nematode feeding, infection by plant pathogens and weed competition) in order to make more accurate predictions.

- Identify, map, isolate (from other organisms) or construct genes to be both transferred into the host plant and expressed effectively to provide improved resistance of corn and soybeans to pests.

- Create sophisticated mathematical models for use in predicting the practical impact of various biological control techniques.

- Develop management approaches to enhance the effectiveness of biological control agents (for both naturally occurring agents and those which must be applied/introduced annually) under various environmental conditions.

- Find more successful approaches of introducing exotic natural enemies, such as pathogens and predators of insect pests and weeds, for use in the corn/soybean agroecosystem.

- Improve the nuclear polyhedrosis virus (NPV) by further development as a biological pesticide.

- Find better methods for applying NPV and nematodes as biocontrol agents.

- Improve the data base on insect pathology to provide a sound basis for developing innovative biological control products.

- Intensify research on combining beneficial insects with insect pathogens for managing insects.

- Identify and synthesize new pheromones and kairomones for use in IPM to enhance the effectiveness of natural enemies including predators of weeds.

- Streamline and improve the review and registration processes for new biological pesticides and biological control practices.

- Define the role of secondary plant chemicals in mediating pest behavior.

Plant Genetic Resistance

Emphasis on classical plant breeding has diminished slightly over the past decade. Molecular techniques will enhance the work of classical breeders and ensure continued emphasis on this important management strategy. Anticipated results from this interaction are as follows:

- More information on plant-pest interactions will be generated at the whole plant, plant pest, genetic and molecular levels.

- Research on mechanisms of plant disease susceptibility and resistance and pest virulence/avirulence will lead to identified genes which could be used to develop plant cultivars with genetic controls for pests.

- The integration of classical breeding with the techniques of molecular biology will open whole new avenues of development guided by such technologies as RFLP mapping and sequencing.

- Existing genetic material will be better characterized, catalogued and preserved for use now and in the future.

Genetic Engineering

The importance of molecular (biotechnology) technologies such as the genetic engineering of plants and microbes through recombinant technology will increase in the future. The technology should not be viewed as a cure-all, but its promise can be improved via the following actions:

- Integration of biotechnology with other research programs dedicated to the development of new agricultural products.

- Development of more cooperation and collaborative funding opportunities among public and private research programs.

- Conduct practical field research to predict how various genetically engineered organisms will adapt in the environment, and that answer basic questions such as "will pest resistance to introduced genes develop, and can it be managed successfully?"

- Fund multidisciplinary teams that integrate a range of technologies from the latest in genetic engineering to long standing agricultural practices.

- Provide education that enhances public awareness of and encourages acceptance and support for these new technologies in food production.

Decision Making and Implementation

Economic injury levels and economic thresholds may differ if biologically enhanced methods are recommended. More research is needed to develop economic injury levels and economic thresholds for many insects and other invertebrates (e.g., mites), diseases, vertebrates and weeds. Fungi, viruses, *B. thuringiensis* and other microbials kill more slowly than do chemical insecticides which could alter existing decision rules. For vertebrates, in particular, this research must be conducted on-farm because population densities cannot be estimated and/or manipulated in small-plot studies typical of research with many other pest species.

- Research and develop the application of economic injury levels and economic thresholds for weed management in

48

corn and soybeans, with appropriate consideration of the soil seed bank and soil microbiota.

- Refine and/or develop additional models for forecasting the occurrence of pest populations, including more widespread use of USDA-Animal and Plant Health Inspection Service survey data.

- Develop optimal sampling plans for many insects, weeds and diseases, as well as meteorological data, to provide real-time data for pest management decisions.

- Collect basic biological data to describe and understand how pest populations are spatially distributed, especially for weed seed banks and seed dormancy.

- Time management strategies in more biologically intensive IPM systems to serve as additional tools for helping growers consultants and IPM advisors integrate all available information into the decision-making process. Providing these tools will require additional systems-level, holistic research and extension programs, and interactive, computer-based delivery systems like computer dial-up and broadcast networks.

- Create an infrastructure to bring together experts from diverse backgrounds to design protocols or guidelines for comprehensive knowledge intensive and biologically intensive management systems.

Pest Resistance Management

The present and future needs in resistance management will require the use of presently available techniques, and those that will become available. A partial list follows:

- Educate extension personnel, consultants and growers on the availability and importance of using the existing resistance management tools.

- Do more interdisciplinary research to better integrate the existing resistance management tools.

- Encourage greater use of currently known parasites and predators, as well as do more research on finding, multiplying, maintaining and releasing parasites, predators and pathogens.

- Research, develop and market more microbial pesticides.

- Research, develop and utilize new and better cultural methods.

- Research, develop and commercialize newer chemicals or biochemicals with multiple modes of action or sites of activity.

- Generate more information on the economics of various cropping systems, IPM programs and the alternating of pest management programs and conventional programs.

- Establish more industry, government, grower and non-profit organization consortia to study resistance issues and viable options.

- Better train personnel to work on resistance management problems.

- Do more research on resistance mechanisms.

- Develop better scouting and monitoring programs to minimize resistance and/or detect it earlier.

- Discover and utilize more pest resistant plants.

- Develop pest population dynamics models incorporating fitness, gene flow and gene expression to describe biological and ecological processes that drive the development and potential management of resistance.

- Apply all available technologies to develop alternative pest management strategies and enhance existing strategies.

IPM Technology Transfer and Extension

Extension should continue to take a leadership role in implementing networks for transfer of new technologies which will become increasingly important as diverse management strategies are developed, tested and made available. Greater support of extension programs will be required to achieve this objective.

- Do more adaptive research focusing on the development of biologically enhanced strategies to support extension IPM programs and outreach.

- Conduct research on IPM implementation including: less labor intensive scouting and monitoring techniques, multiple pest thresholds which take into account species complexes that occur simultaneously or sequentially, data on the effect of cropping systems on pest and beneficial species, data to improve the knowledge base of host development and pest interaction to quantify potential loss and yield responses, and a better understanding of pest biology which identifies potentially unique vulnerabilities that might be exploited in IPM programs.

- Hire more CES specialists to develop, demonstrate and deliver programs in the more labor-intensive area of biocontrol. Expand and refine ongoing programs and develop new ones, including those directed toward different rotational and tillage systems and levels of management and inputs.

- Direct more IPM training toward growers.

- Use interdisciplinary teams more effectively to address problems, and to develop and transfer pest management strategies including CES, researchers, consultants, growers and those in agribusiness.

- Increase communication networks among the states for more interactive cross communication among researchers and extension specialists, and for maximum mutual understanding of the roles each play in IPM.

- Integrate pest management tactics into a holistic crop management system.

- Do more on-farm demonstrations of scouting and monitoring techniques, simplify the steps in the pest management decision making process, and encourage the application of biologically enhanced management strategies.

51

- Determine environmental short-comings for each biologically enhanced strategy.

- Do surveys on economic, environmental and social impacts of programs on clientele needs.

- Encourage CES specialists to provide more intensive education for CES agents.

- Expect CES agents to coordinate on-farm demonstrations, conduct local training and catalyze local IPM adoption.

- Develop expert logic systems for use by producers and other agribusiness community.

- Initiate a program designed to work with private practitioners who presently deliver IPM technology.

- Exchange information among private consultants who are successfully increasing IPM use by growers.

Agricultural Chemicals

Agricultural chemicals are now and will continue to be key elements in IPM programs for the corn/soybean agroecosystem. To support the safe use of present pesticides and to encourage the development and registration of new products, research on interactions between chemicals and the environment should be encouraged. A better understanding of potential risks and benefits will allow future pesticides to be developed and used in a manner which will reduce unanticipated effects while enhancing their effectiveness in controlling targeted pests. This can best be accomplished by:

- Funding programs that evaluate compatibility of existing and new pesticides with IPM systems.

- Educate growers in the proper use of pesticide products within IPM programs.

Regulations

To establish IPM as a key component of the long-term goals governing institutional policies set by regulatory agencies and legislative bodies, close coordination is necessary among regulators, policy formulators, researchers and practitioners. Two key elements will ultimately determine the success of IPM and its components, 1) the seller (consultant) who delivers the service and 2) the buyer (grower) who uses the service. The service must be profitable for the consultant, and the grower must experience economic gain. Input is needed at all levels from both the public and private sectors. The following actions are suggested:

- Review the present federal and state production and support programs, and suggest changes to protect the grower who adopts new IPM technology and practices.

- Review intensively and streamline the entire pesticide regulatory process, especially for the regulation of organisms used for biological control, registrations of minor and rotational crop uses of pesticides and registrations of new, more selective chemical and biologically active ingredients.

- Develop a certification program for IPM advisers and/or support existing programs such as Registry of Environmental Agricultural Professionals (REAP) and National Alliance of Independent Crop Consultants (NAICC).

- Emphasize practical regional and state regulatory requirements.

- Coordinate state, federal and international product registration requirements for compatibility.

- Complete the rule making process for the registration of biopesticides and products containing genetically engineered organisms.

CONCLUSION

The corn/soybean crop production system concentrated in the 10-state Corn Belt region is the largest agroecosystem in the United States, and perhaps the most finely tuned in the world. However, there are constraints to the development of an IPM system. These constraints include:

- Low cash value crops

- System-level applied research

- Funding

- Lack of personnel who are integrating the various factors of the system for more site specific management

Although this agroecosystem accounts for nearly half of the $5.3 billion expended annually in the United States for pesticides, they are used judiciously. Corn/soybean production systems fully utilize available commercial sources of host plant genetic resistance against plant pathogens, nematodes and insects. The important question is how sustainable is this system, and what is the potential now to become even more sustainable? Can its sustainability be improved through the adaptation of biologically enhanced IPM programs? Succinctly stated - "if it pays, it stays."

Weed control is the biggest problem facing producers in the corn/soybean agroecosystem. Synthetic chemical herbicides continue to be the most economical and effective method of weed management. This is due, in part, to lack of alternative methods. Little IPM effort has been spent on weeds compared with insects and diseases. A high potential exists to reduce, but not eliminate, herbicide and insecticide use, and to minimize the development of pest resistance in corn/soybean agroecosystems, including those of continuous corn production, through adoption of IPM. But this potential cannot be realized simply by implementing current technology. Additional research is needed. Chemical pesticides will continue to be an important pest management strategy for many years. However, biologically enhanced management systems are already in limited use with great potential for additional implementation. The question is not will it happen, but how rapidly will we move away from our heavy reliance on chemical pesticides? The time required for expanded adaptation is governed by economic forces, and the availability of appropriate technology implemented at the grower level.

By addressing the constraints raised in this report, policymakers and ultimately producers can move more quickly towards a sustainable system with maximum use of biologicals. Resolving these constraints will require a coordinated effort to influence both the public and private sectors.

CORN/SOYBEAN IPM ACTION TEAM

The following individuals were committee members of the Corn/Soybean IPM Action Team, and contributed greatly to the development and review of this report.

Dr. R.E. Ford (Co-Chair)
Department of Plant Pathology
University of Illinois

Dr. C.R. Edwards (Co-Chair)
Department of Entomology
Purdue University

Dr. Leslie C. Lewis
USDA-ARS
Corn Insects Research Unit
Ankeny, IA 50021

Dr. Marlin K. Bergman
Pioneer Hi-Bred International,
Inc.
Johnston, IA

Mr. A.L. Christy
Crop Genetics International
Hanover, MD

Dr. L.T. Hargett
Sandoz Crop Protection
Des Plaines, IL

Dr. Rhonda Janke
Rodale Research Center
Kurztown, PA

Dr. John Cardina
Department of Agronomy
Ohio State University

Mr. Mike Erker
National Corn Growers
Association
St. Louis, MO

Dr. Elizabeth Owens
ISK Biotech Corporation
Mertor, OH

Dr. Dennis Keeney
Leopold Center
Iowa State University

Dr. Keith Smith
American Soybean Association

Mr. Dave Harms
Crop Protech
Naperville, IL

Dr. Frank Serdy
Monsanto
St. Louis, MO

Dr. William Ruesink
Illinois Natural History Survey
Champaign, IL

Dr. Lowell Getz
Department of Ecology,
Ethology and Evolution
University of Illinois

Dr. John Dudley
Department of Agronomy
University of Illinois

Chapter 3

BIOLOGICALLY INTENSIVE INTEGRATED PEST MANAGEMENT: FUTURE CHOICES FOR COTTON

Raymond E. Frisbie, Director
Center for Biologically Intensive
Integrated Pest Management
Texas A&M University
College Station, TX 77843

D.D. Hardee, Laboratory Director
Southern Insect Management Laboratory
USDA-ARS
Stoneville, MS 36849

L.T. Wilson, Professor
Department of Entomology
Texas A&M University
College Station, TX 77843

The management of pests has been an integral part of the evolution of the cotton industry in the United States. Particularly since the turn of the century, the quest for solutions to pest problems has been a dominating concern for this high-value crop. Perhaps more than any other crop, cotton has been central to the development of integrated pest management (IPM) as a science and a philosophy. Intense competition for resources between the plant and the myriad of pests affecting the plant has focused the attention of scientists from every discipline within agricultural sciences and there has been progress. Scientists from the Land Grant universities, The United States Department of Agriculture (USDA), and private industry have been successful at developing strategies and techniques to control arthropod, plant pathogen, nematode, and weed pests. Although significant gains in managing individual pest species have been made, scientists have only begun within the last decade to view cotton pest management within the context of the total production system of cotton and the surrounding agricultural community that contains it.

Cotton production is a highly complex biological, physical, mechanical, economical and political system in which pests play a significant role. Cotton has been one of the major agricultural crops grown in the United States since the late 1700's (9) and currently ranks as the fourth most valuable crop in the United States after corn, soybean, and wheat. A total of 11.5 million acres of cotton were harvested in 1990 with an average yield of 640 lb lint per acre. Texas, California, and Mississippi were the three leading states in production and accounted for almost 60% of the total crop (10). The estimated value of the 1990 crop from 15 cotton producing states was near $4 billion for lint, with an additional $507 million for cottonseed (2). The management of weeds, insects, plant pathogens, and nematodes is a major component of cotton production. In 1990, losses to these pests were estimated to be $1.03 billion (14).

The cotton industry has undergone notable changes in the last 15 years, and as we look to the future, will undergo even more changes by the end of this century. For example, there is a distinct trend for cotton to be produced on larger farms (16). The market share of farms with sales greater than $500,000 increased from less than 7% in 1969 to 48% in 1982 (16). In the same period sales from small farms ($20,000 to $99,000) declined from 56% to 14% of the market. Change in farm size will directly affect management decisions and must be taken into consideration in designing IPM systems. In some ways larger farms may facilitate the management of pests by allowing a more centralized, homogeneous approach to pest management (5).

Given shifts in technology, farm size, economics and policy affecting United States cotton production, the Cotton Action Team: 1) Surveyed the cotton industry to assess current technology and identify major constraints to IPM research and implementation in the United States, and 2) Engaged in an interactive workshop (Wingspread Conference) to develop a blueprint for future cotton IPM systems that considered research and education priorities for both public and private sectors.

A SURVEY TO ASSESS IPM PRIORITIES AND CONSTRAINTS

A survey (see Appendix A) of over 60 of our cotton IPM colleagues across the cotton belt was recently conducted. The survey identified research and extension needs for development of biologically intensive IPM tactics, elaborated on the constraints to their usage, and determined what could be done to reduce or eliminate such constraints. Biologically intensive IPM for the purpose of this survey was defined as the use of biological control, host plant resistance, cultural management, and judicious use of environmentally safe pesticides for the control of multiple pests. In addition to the survey results, pertinent literature was also reviewed and included in this report.

Our intention is not to suggest that pesticides do not work and are not solving pest problems. On the contrary, they do work well in most cases, and that is the main reason they are widely used. However, the development of pesticide resistance in pests, concerns over environmental issues and human health, lack of alternative management practices, the emergence of more pests species, i.e., the cotton aphid and the sweetpotato whitefly, and an ever-shrinking arsenal of chemicals available for cotton pest management underscore the need to develop a broader array of control tactics.

AVAILABLE COTTON IPM TECHNOLOGY

Currently, there are a variety of tactics used or that show potential for use in IPM programs. A major challenge still exists in adapting and using these tactics for commercial cotton production. Some of these tactics include:

- Cultural control (crop rotations, cover crops, and conservation tillage systems that minimize soil disturbances to promote allelopathic effects and suppression of weed populations; timely crop termination, early stalk destruction and antagonistic cover crops, and management of overwintering habitat; soil solarization, optimum seeding rates of high quality seed, planting at near optimum soil temperature on raised beds, use of pest resistant cultivars, and the use of proper nitrogen rates and irrigation practices to minimize plant pathogen damage),

- Use of Low Energy Precision Application (LEPA) systems for irrigation and chemigation (alternate row, high frequency, and deficit irrigation systems may reduce incidence of weeds and diseases and greatly enhance water and soil conservation; chemigation directs pesticides to target zones on plants where damaging infestations occur; fertilization

59

with deficit irrigation limits movement of soluble fertilizer elements),

- Host plant resistance (genetic resistance, genetic tolerance, morphological traits, i.e., nectariless, frego bract, okra-shaped leaf, glabrousness, and high gossypol and tannin),

- Genetic engineering using biotechnology (herbicide resistance [glyphosate, bromoxynil and 2,4-D] and insect resistance using *Bacillus thuringiensis*),

- Use of trap crops as refugia for natural enemies and/or selective spraying of pests to preserve natural enemies,

- Biological control with parasites, predators, pathogens and nematodes (conservation, inoculation, augmentation, habitat management),

- Bioinsecticides (viruses, *Bt*, IGRs) and biofungicides,

- Use of semiochemicals for population management and in survey traps for timing pesticide applications,

- Use of sterile insect releases (e.g., pink bollworm in California),

- Increased use of crop consultants and demonstration programs by extension personnel,

- Increased tactical use of regional programs rather than individual farms, and

- Increased use of systems management tools such as simulation models, and integrated expert systems to bring about a more objective and consistent management of individual farms and whole regions.

IPM CONSTRAINTS

Major constraints which might prohibit the development and/or use of the above tactics include:

- Grower attitudes (perceived economic risks, skepticism and tradition, difficulty of administering, convenience, effectiveness and cheapness of pesticides, reluctance to risk maximum yields and requirement of more training and skill),

- Lack of extension education and demonstration programs on effectiveness of IPM tactics,

- Inconsistency or inadequacy of pest control with current biopesticides,

- Inadequate financial support of breeding programs and biological control programs,

- No incentive program or legislation for IPM use,

- High capital demands for land use which encourages the concept of producing maximum yield rather than optimum and most profitable yield, and

- In some states, such as California, rigid policies of state pesticide regulatory agencies can impede the implementation of IPM tactics, e.g., soil treatment for plant pathogen and insect control, designed to reduce overall pesticide use.

Some of the current state/federal policies and/or regulations which constrain the use of biologically intensive IPM tactics and their solutions include:

- Both USDA and the Environmental Protection Agency (EPA) strongly support research and implementation of IPM. USDA is supportive for reasons of increased profitability for American agriculture and for reducing non-target effects of pesticides. EPA is supportive of IPM as a way to reduce or eliminate pesticide pollution in agriculture. Despite stated policies of both agencies, funding is often inadequate for IPM research and extension programs. Lack of sufficient funding is singularly the largest constraint in the development and implementation of IPM. To overcome this constraint, it is recommended that IPM research and extension funding be substantially increased in order to align funding levels with stated agency policies.

- The provisions of the Farm Bill encourage growers to strive for high yields in order to make the most of subsidy payments. The incentive for higher yields to increase deficiency payments may cause farmers to use greater inputs, such as pesticides. It is counter intuitive to ask farmers to accept lower yields unless it can be demonstrated that net profits can be increased or unless the provisions of the Farm Bill are such that farmers are rewarded for producing commodities under IPM. To address this constraint, it is not unreasonable to change the Farm Bill to reward farmers for practicing IPM through agricultural loan procedures, tax credits, federal crop insurance incentives, more flexibility for crop rotation or by awarding additional base acreage.

- Universities and the USDA have developed a general policy to shift toward more basic research, which may neglect applied ecological and practical field studies needed to develop and refine IPM techniques. In order to address this constraint, it is recommended that a more balanced policy which more fully integrates basic with applied ecological research be pursued by the universities and USDA. Further, a policy of extension being more involved in designing and participating in applied research should be implemented.

- The general policy and structure of universities and USDA is one of separate pest and production disciplines. Although great strides have been made to develop interdisciplinary, systems-oriented approaches to IPM, more emphasis is required that binds IPM techniques, production system tactics and economic considerations together into the total crop system.

- A major constraint exists with the pesticide review and registration process for pesticides by EPA. This process is time consuming and costly for the registrant. Streamlining of the registration process to improve efficiency is recommended. Definite EPA requirements and guidelines are necessary for the registration of transgenic plants and microbes. Further, EPA should facilitate the registration process for IPM-compatible products, such as semiochemicals, microbials and transgenic plants. The EPA registration and reregistration process should consider the impact of new products and removal of old products on existing IPM systems.

- Because of the great diversity of American agriculture, research and implementation of policies and regulations should become more regionally specific. To facilitate regional specificity it is recommended that Hatch, Smith-Lever and other federal funds be provided to the Land Grant universities for establishing regional and subregional Centers of Excellence for IPM.

- State policies and regulations should be strengthened and enforced for improved postharvest stalk destruction to minimize overwintering insect populations and to support, where appropriate, better plant pathogen and weed management practices.

- As greater emphasis is placed upon IPM, particularly biologically intensive IPM, university policies in the future should be strengthened and incentives provided to recruit top quality students that are trained to work with field-oriented IPM programs.

- Policies should be considered that allow the supporting or subsidizing of biologically intensive IPM programs in environmentally sensitive areas. For example, the Endangered Species Act may prevent the use of pesticides in certain areas. In order for production to continue, biologically intensive IPM systems should be developed to allow for economic crop production.

- Soil conservation policies may conflict with water conservation and pesticide pollution prevention policies. Soil conservation requires maintaining a crop residue on the soil surface in some areas. These residues may serve to increase insect and plant pathogen populations. On the other hand, this practice may serve to reduce weed populations. Environmental policies from the USDA Soil Conservation Service (SCS) and EPA should be evaluated and made compatible where appropriate.

CHALLENGES AND OPPORTUNITIES FOR COTTON IPM IN THE FUTURE

Several challenges were identified by those surveyed. These challenges ranged from increased funding for IPM, to more emphasis on biological tactics, to major institutional and policy shifts.
- The primary challenge facing cotton IPM in the next several years is clearly the establishment of a stable and long-term

63

funding base for research and extension programs. Present political, economical, social and environmental issues will continue to pressure cotton and other crop producers to consider using less pesticides. In the absence of viable economically and environmentally safe alternatives, there is an urgent need to increase federal, state and private funding for expanding IPM programs. Without expanded funding, research and extension will not remain productive in providing adequate pest management strategies for the United States cotton industry.

- New approaches by research and extension scientists are necessary to explore creative and non-conventional strategies for pest control. These approaches should continue not only to be profitable, but should also be sustainable for the long-term. For the last 45 years, pesticides in general have dominated cotton pest control. Agricultural chemicals will have a valuable role to play in the future, but to a lesser degree with the development of biologically intensive IPM systems that place emphasis on host plant resistance/biotechnology, biological control, and cultural manipulation.

- A major challenge for the future will be the management of entire pest populations on an area-wide basis. Traditionally, IPM programs have been developed for individual farms. Sufficient insect management examples exist showing that greater suppression of total populations is realized and are more economically and environmentally successful when there is area-wide participation of farmers. Examples, such as short-season cotton IPM systems, stalk destruction, uniform planting dates, sterile pink bollworm releases in the West, boll weevil fall control (diapause) programs, early season boll weevil control programs, community wide bollworm management, tobacco budworm resistance management, and the boll weevil eradication program show that populations can be managed more efficiently on an area-wide basis. Area-wide suppression programs are not meant to replace individual, farm-level IPM programs, but rather to serve as regional or area-wide complements of farm level programs.

- The ever pressing challenge of gaining farmer acceptance of biologically intensive IPM techniques will continue. The burden, first on the extension-scientific community and then on researchers, will be to develop and widely test IPM techniques that are economically acceptable to farmers. Much more emphasis will be placed on efficient management which will place even more of a burden on extension educators. A need exists to establish research validation farms, such as AG-CARES in the Texas High Plains, to demonstrate the effectiveness and practicality of adopting new crop management systems.

- Critical for the future is expanded curricula and incentives to attract talented students into IPM programs. As agricultural enrollment continues to decline within the Land Grant Universities, innovative ways of attracting students must be developed in order to have a constant supply of IPM experts for the future.

- Future cotton cultivars with favorable yield and quality characteristics must be bred with multiple pest and multiple stress resistance. The role of biotechnology, with all of its powerful potential, must be more fully integrated into major breeding programs. This will require the research forces of public and private cotton breeding programs to be more actively merged.

- The potential for a decreasing availability of pesticides will place a short-term stress on the cotton industry. Pesticide resistance only exacerbates this situation as fewer pesticides are available and greater selection pressure for resistance is placed on remaining pesticides, thus creating a vicious circle. The long-term solution for pesticide resistance, particularly of the tobacco budworm, will not be found in the development of new insecticides or the mixing and rotating of existing insecticides. Resistance management for this insect will rely on other factors, such as cultural controls, to achieve sustainable management.

- The agricultural chemical industry will be faced with the challenge of developing cost effective, environmentally safe pesticides with a low probability of pest resistance development in the future. This, perhaps, is the tallest order of all. Expanding the range of pesticides available through biopesticides and transgenic plants can supplement current chemical insecticides facing resistance problems. However, even with emphasis on developing and engineering improved strains of B. *thuringiensis* through transgenic plants or microbial insecticides, the specter of resistance is still on the horizon. Resistance management programs for new pesticides will have to be developed. Again, with tobacco budworm, resistance development is almost certain. Current and future insecticides must be looked upon as a viable resource to be protected.

• The introduction of new pests or the elevation of minor pests to key pest status will be a constant challenge in the future. For example, the sweetpotato whitefly, a major pest in the West is causing significant problems in South Texas. The cotton aphid, previously considered a minor pest, has become a key pest over the last five years. The morning glory complex is an escalating weed problem across the cotton belt. Silverleaf nightshade in West Texas has increased in intensity over the last several years.

• In a broader context, there are policy questions that should be addressed relative to cotton production: Which geographic areas of the United States are most suitable for profitable and environmentally acceptable methods of cotton production? Are there certain considerations that might rule out production in specific regions or localities? Are there some environmental considerations (infertile soil, lack of water, human population density, pest populations) that are contradictive to cotton production? What crops could be substituted for cotton in those areas where continued monoculture production of cotton is counter-productive or environmentally unsuitable? Can some cotton acres be converted to non-agricultural use, such as conservation or development, and how will farmers be compensated?

COTTON IPM: A BLUEPRINT FOR THE FUTURE

Members of the Cotton Action Team met for a structured, interactive work session during a forum sponsored Wingspread Conference. Given a broad based, multi-disciplinary historical perspective and the above survey results, the team developed an initial working blueprint for research and educational programs for cotton IPM in the future. Much of the Cotton Action Team's findings directly support, complement, and in some cases expand upon the research and education priorities identified in the survey. This blueprint was widely reviewed by all sectors of the cotton industry.

The Importance of Scale

Cotton IPM systems must consider not only individual elements within the system in relation to pests, production practices, economics and the environment, but they must also consider the scale or size of programs to be implemented. Classically, but with notable exceptions that will be discussed later, IPM systems have been developed for the individual field or farm. Without a doubt, individual, farm-level IPM programs have been and will continue to be a focal point in managing pest populations. However, owing to high levels of pest mobility, pests should be considered "common property." What an individual farmer does or does not do relative to the management of pests directly or indirectly affects his neighbors and consequently the entire production region. Hence, producers share a common responsibility in pest control. This being the case, cotton IPM programs of the future will consider developing technology not only at the farm level, but will give added attention to the regional (communities within states or clearly identified production regions) or even at a larger, multi-regional scale (groups of regions, states or national).

Figure 1 identifies insect IPM program examples at the farm-field, regional, and multi-regional levels. Similar examples may be developed for plant pathogen and weed management programs. IPM tactics must be developed that are size sensitive, and dependent on the level of scale being addressed. The kind of research devoted to the various elements of the developing IPM system will by necessity take on a different focus. For example, a field-level IPM program being developed for an individual farmer that considers crop rotations, cultivar selection, cultural practices, scouting and treatment threshold procedures, as well as choice of insecticide, might be substantially different from a regional IPM program. A regional IPM program would require specific practices of individual farmers to be coordinated, such as planting dates and related cultural practices (stalk destruction as an example), sampling techniques that would provide regional indices of pest populations, coordinated pesticide application decision methodologies and perhaps a strategy for a crop rotation that would generally benefit the region. Additionally, educational techniques for regional programs would differ from individual farmer or field-level IPM programs. Legal considerations, such as pink bollworm and boll weevil laws that have planting and stalk destructions mandates, should be weighed against voluntary programs. Area wide IPM programs require the involvement of local, regional, state or national producer organizations (or the formation of producer foundations as

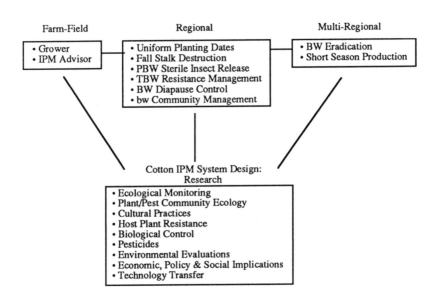

PBW - Pink bollworm
TBW - Tobacco budworm
BW - Boll weevil
bw - Bollworm

Figure 1. Range of Scales for Managing Insects in Cotton IPM.

in the boll weevil eradication program), supported by legislation, to gain regional and multi-regional agreement and cooperation among farmers.

Regional and multi-regional programs have shown to be more cost effective for the individual producer for certain pest species. Boll weevil, bollworm, and pink bollworm management programs, as well as tobacco budworm resistant management programs have been significantly more effective in dealing with pests on a regional scale when compared with the farm-field level of scale. Dr. E. F. Knipling, long an advocate of total population management at the regional and multi-regional level, addressed the problems and promises of total population management in his response to the survey mentioned in the first section of this report. He states:

"It will be no simple matter to make the transition from control measures applied by growers at their discretion, to coordinated programs executed by pest control agencies. This would have to be recognized by agricultural administrators, appropriations committees, the cotton industry and others. However, if the industry is indeed threatened by the lack of satisfactory pesticides and the public becomes increasingly apprehensive over environmental hazards created by pesticides, our scientists must give serious consideration to drastic changes in strategies, as well as in the techniques of control. For the major cotton insect pests the total population management system, in my view, offers the only hope that cotton can be produced with little or no reliance on insecticides. The applications of insecticides is now the chief component of IPM programs. If future control programs are intended for voluntary execution by individual growers, the use of insecticides will almost certainly continue to be an essential component. In such an event, the major thrust in research should be to discover effective and environmentally acceptable insecticides. Judging from the successes during recent years, the chances of finding new, safe insecticides may not be too good."

Knipling goes on to say:

"To make the transition from voluntary programs to be executed by growers, to the total management systems to be executed by pest management organizations, I envision two phases. The first will be to make use of all available control procedures in a fully organized manner to reduce the generally high populations to very low levels. This may require, for one or two reasons, the intensive use of insecticides, rigid cultural practices, growing resistant cultivars and any other practical methods. The greatly reduced populations would then be maintained year after year at sub-economic levels throughout ecosystems by such techniques as the routine release of genetically altered insects, the release of mass-produced parasites or other biological agents, and the use of attractant traps or attractant bait stations. Combination of two or more such techniques may be advantageous. These techniques are all highly pest specific so that maximum help should be gained from natural biological agents. Maximum help from natural agents is one of the objectives of IPM programs, but since most programs still rely largely on non-specific insecticides this objective is not now achieved" (E.F. Knipling, *personal communication*).

Elements of the Blueprint

Given the importance of the scale or level of program emphasis, the Cotton Action Team members were asked to design a cotton IPM system for the future. Having identified primary policy and institutional constraints in the previously discussed survey, the blueprint for the future focuses mainly on the technical and educational opportunities for cotton IPM over the next 10-15 years. This activity required that components of the IPM system be examined separately and as they relate one with another. As noted in Figure 1, the eventual system, or design, that emerged considered the following elements: ecological monitoring, plant/pest community ecology, alternative cultural practices, host plant resistance/biotechnology, biological control, pesticides, environmental evaluation of the emerging IPM system, economic, policy and social implications, and innovative educational and technology transfer technologies. These elements are reviewed in the following sections. Priority future research and education needs are identified.

Ecological Monitoring

Monitoring techniques for plant development, pests and natural enemies involving sampling decision rules for improved methods of quantifying pest abundance or potential crop loss can significantly reduce the risk of pest control, optimize the use of pesticides, and improve adoption of alternative practices. In addition, sampling techniques for environmentally sensitive, non-target organisms, water, soil and air can be used or developed to provide a parallel environmental assessment of developing systems. Treatment or action thresholds developed should be based on the relationship of pest occurrence or abundance to measurable crop loss in order to insure cost-effective environmental decisions.

Expanded monitoring research is needed for weed management. More information is needed on the severity and loss caused by individual weed species and complexes of weeds. The frequency, intensity and reliability of weed sampling techniques should be determined. The benefit of weed monitoring should be evaluated and the possibility of incorporating weed sampling and thresholds into crop monitoring services should be investigated.

Monitoring needs for plant pathogens are broad and far reaching. There is need for basic research to provide estimates of pathogen abundance related to etiology, epidemiology, pathogen variability, and host-pathogen-environment interactions. There is a need to determine the relationship of pathogen inoculum density to plant infection for most of the major pathogens attacking cotton.

Despite the tremendous work done in developing monitoring techniques for insects, there remains much work to be done both for farm level and regional management decisions. Cost-benefit analysis should be expanded for all sampling techniques in order to determine their practical utility under field conditions. Relative to cost effectiveness, sampling techniques of the future should be simple and easy to use by the farmer or crop consultant; for example, the number of nodes above white bloom as an indicator of plant cutout. Thresholds should be stratified to consider plant stage vulnerability. Natural enemies should be incorporated into sampling and threshold decision making in a reliable, workable way. There is a need to standardize survey and sampling procedures in order to insure comparability between areas being sampled. Sampling techniques for new pests, such as the sweetpotato whitefly, should be developed that detect these new pests at low levels before larger acreages are impacted. Multiple-species thresholds should be developed to consider the range of insect pests that attack cotton simultaneously. Multiple year thresholds that consider measuring population changes over longer time periods should be constructed to help in regional and farm level pest management decision making. Sampling for insecticide resistant species should be emphasized in order to provide an early warning of increasing resistance within insect populations.

Plant/Pest Community Ecology

Whether the scale of emphasis is the individual field or a national eradication program, research on fundamental plant/pest ecology should focus on quantifying when and how pests affect crop growth, quality and yield. Several cotton plant and pest models have been developed in order to gain a better understanding of complex plant/pest relationships. Work in this area should be expanded to quantify the basic biological and ecological processes and relationships between plant-pest-natural enemy trophic levels. Studies should be conducted that address the impact of changing management approaches upon the interaction of pest with their host plant, habitat, community or ecosystem in order to lead to a substantial improvement of our knowledge on the development of biologically based management recommendations.

In order to make simulation models more workable at the field or tactical level, simple, time efficient sampling techniques should be

71

developed that provide the basic data in order to drive models. Expanded emphasis should be placed on developing both strategic and tactical models for weeds, nematodes and plant pathogens. These models should be linked with plant models. Multiple-pest models, within and between pest classes, should be developed. Models should include economic elements that place a value on yield and pest induced crop loss. Farm policy and economic models must be constructed that help evaluate IPM options for individual farmers and for larger, regional programs. For example, changes in farm policy may facilitate or hamper area-wide IPM programs.

There should be an expanded multidisciplinary, team approach to constructing models. Models can serve as an appropriate organizational framework for "integrating" disciplines representing various biological and production components of an IPM system. Further, there should be expanded collaboration, e.g., university-USDA, in the development of models in order to reduce redundancy in model development and increase portability of models to various geographic areas within the cotton belt.

Cultural Practices

Cultural manipulation provides the essential structure for biologically based alternatives for the management of pests. Cotton has several excellent examples of using cultural practices as the primary base for IPM programs. These include rotations, cultivar selection, planting or harvest dates, fertility and water management, and phytosanitation practices. Research in this area should not only consider reduced pest damage, but should also be linked with research aimed at reducing soil erosion and environmental contamination by fertilizer.

Specifically, expanded investigations should be made to take advantage of allelopathy for weed control. Isolation and synthesis of allelopathic compounds or organisms may provide new frontiers in weed management. The incorporation of intercropping or cover-cropping systems to suppress weeds should be linked to conservation tillage systems. Planting patterns or planting geometry should be investigated to determine optimum weed suppression. The role of living or sprayable mulches, particularly for weed, nematode, and plant pathogen management, should be examined. Companion plantings may be useful as trap crops for insects or serve as refugia for natural enemies. Companion plants that serve as a between season biological bridge for natural enemies should be investigated. For example, in production regions with mild winters, the sweetpotato whitefly has many parasites that could increase in population throughout the year, outside of the cotton field, in a

72

companion planting of sunflower. The population suppressive value of a host free period should be reinforced through well planned research. As alternative cultural systems are researched, they should be compared with conventional systems in terms of productivity, net profit and environmental impact. This of course is true with any new IPM tactic, whether it is a cultural alternative or not.

Host Plant Resistance/Biotechnology

Maintaining plant health is essential for profitable cotton production. Resistant cultivars combined with cultural management tactics have and will continue to provide a strong, ecologically sound foundation for cotton IPM systems in the future. Evaluations of new plant materials or cultivars should focus beyond the target species. The impact of nontarget pest species and natural enemies must be assessed during the evaluation process.

Research incorporating multiple resistance traits including insect, plant pathogen, nematode and environmental stresses should be pursued in order to achieve the true genetic potential of the cotton plant. In addition, the cotton plant of the future must be highly efficient in its use of solar radiation, water and nutrients in both irrigated and dryland production.

Available insect resistance traits, such as nectariless and glabrous for bollworm/tobacco budworm resistance should be more broadly incorporated into commercial cultivars. The search for sources of pink bollworm, cotton fleahopper and plant bug resistance should be expanded. Research on quantifying the value of short season cotton cultivars must be strengthened.

The field of biotechnology offers opportunities to enhance efforts in developing new cotton cultivars. New molecular genetic technology provides the prospect for conferring resistance to insect and plant pathogens, and tolerance to herbicides, through the introduction of genetic material from sources whose resistance cannot be incorporated through sexual hybridization. For example, the insecticidal gene for *B. thuringiensis* has been added to cotton and will soon be available in a commercial cultivar. This resistance trait, as well as others developed through other plant breeding techniques must be incorporated into a balanced, multi-tactic IPM program in order to prevent pest resistance to these valuable traits. Additionally, biochemical means for identifying resistance genes should be developed. Finally, research should be expanded to determine the true risk, if any, of introducing genetically altered material into the environment, i.e., the products of biotechnology.

The field of simulation modeling should be applied to cotton breeding. The use of modeling will enhance efforts in identifying useful traits and provide additional decision aids to genetic improvement programs. Simulation modeling could be used as a tool to evaluate strategies for the effective use of herbicide resistant and bollworm/tobacco budworm *B. thuringiensis* resistant cultivars.

Stable and sustained financial support is necessary for genetic improvement programs to be productive and to continue developing superior cotton germplasm and cultivars. Historically, cotton breeding programs have given the United States cotton producers a globally competitive advantage. Support for genetic improvement research cannot be turned off and on at will if productive programs are to be maintained. Collaboration between public and private breeders must be strengthened in order to deliver cotton cultivars with good yield, high quality and multiple pest resistance.

Biological Control

Biological control should receive major emphasis in future IPM programs. As cultural control alternatives are developed along with pest resistant cultivars, the impact of these tactics should be carefully considered in light of existing or potential natural enemies.

A major void exists in the biological control of weeds. Weeds present a particular problem owing to the diversity of species that simultaneously compete with cotton. Biological control agents, whether they be insects, nematodes or microbial weed pathogens, are by nature highly specific to individual weed species. Biological control initiatives should be considered not only for individual species, but for complexes of weeds as well. The integration of biological and chemical control should be evaluated for complementarity.

Biological control of plant pathogens should focus on the development of seed treatments. More fundamental research is needed on the mechanisms involved in biological control of plant pathogens, their interaction with abiotic and biotic agents, and the influence of the environment.

The conservation of natural enemies has been a primary focus of biological control of insect pests. Despite a large body of work, it is still necessary to develop life tables to accurately assess the role of parasites and predators in insect suppression. Life tables will also be helpful to incorporate natural enemies into economic threshold decision making. Additionally, the amount of mortality caused by natural enemies must be established before augmentation with mass-reared natural enemies, e.g., *Trichogramma* spp., can be effectively used in management programs.

There should be an accelerated emphasis on identifying and introducing exotic natural enemies of the cotton fleahopper and the plant bug complex. Major initiatives are needed in the biological control of new pests, such as the cotton aphid and the sweetpotato whitefly. Because of the high reproductive rate of these pests, along with their propensity to develop insecticide resistance, a broad based biological control initiative in concert with plant breeding and cultural manipulation initiatives is in order. As exemplified earlier, the role of manipulating alternate hosts of the sweetpotato whitefly to serve as in-season and between season (living bridges) insectaries for parasites should be investigated. Expanded work in identifying and developing pathogens, particularly fungal pathogens, for these two species is needed. Similarly, research on the use of fungal pathogens as potential biological control agents for spider mites is also needed.

Research on mating disruption and sterile male releases for pink bollworm should be broadened. Communitywide management of bollworm and tobacco budworm using viral and bacterial pathogens should be further evaluated with particular emphasis on early spring generations.

Pesticides

There will be a continued need for biologically and chemically based effective, non-disruptive and environmentally safe pesticides. Public and private interests must work together to develop new and varied chemistries, as well as microbial pesticides, pest growth modifiers and other pesticides that fit into balanced cotton IPM systems of the future.

In cotton, strategies to gain the most effective use of pesticides and avoid resistance, particularly with tobacco budworm, cotton aphid, sweetpotato whitefly and other pests, should be developed. The class of pesticide and the timing of their use should be weighed against the pest suppressive value of the pesticide along with the preservation of natural enemies.

The area of effective and safe pesticide application should be elevated to a major research category. Drift control research, including electronic sensor control of application equipment, should be stressed to minimize the impact on off-target organisms, including man. Advanced and safer systems for pesticide mixing and loading to reduce worker exposure should be emphasized. The new post-emergence herbicides offer promising areas of research on the selection, timing and application of these important pesticides.

The role of pesticides in regional cotton IPM programs should be further evaluated. Whether microbial, pheromone or traditional chemical based pesticides are used, the advantages and disadvantages of regional applications should be evaluated.

Environmental Evaluations of Emerging Technologies

Any change in the cotton production system must realistically consider the ecological and economic benefits and costs to society. Because cotton production is a major user of pesticides, future research should focus on a greater understanding of the impact of conventional and alternative pest management strategies on the long-term sustainability of managed ecosystems. Research emphasis may be placed upon providing a quantitative comparison of emerging IPM technologies with those technologies that are subject to replacement or modification. It is in the best interest of the United States cotton industry to place priority on techniques or procedures that provide a rapid evaluation of pest management approaches, such as monitoring or assay methods and the development of pesticide fate, soil level and regional level environmental simulators.

Economic, Policy and Social Implications

Adoption of cotton IPM techniques should change significantly relative costs and returns. Economic policy and risk analysis should be done at the farm-firm level, regional or multi-regional (even national) levels. Expected costs and returns of selected products and inputs of new or improved IPM systems will be used to identify priority areas of high research and technology transfer.

Innovative Educational and Transfer Technologies

Gaining adoption of IPM techniques in cotton will remain a major challenge in the future. Although cotton has a good track record for adopting new IPM technology, research should focus on innovative ways to most efficiently transfer this technology at farm, regional and multi-regional levels to gain adoption in the least amount of time. Greater management and interpersonal skills will be necessary for research, Cooperative Extension Service (CES) and the private sector in order to accelerate adoption of new technology.

Because of the site specificity of pests down to the individual farm, IPM technology transfer will continue to emphasize the training of producers and private consultants in the making of field level decisions. The CES must accelerate and expand training programs to accommodate more producers in the future. Novel teaching methods must be developed to extend the training capabilities of CES IPM specialists. The private sector should be involved in the development of multi-factorial training programs. Both public and private media expertise should be included in the development of innovative technology transfer methods.

Perhaps the greatest challenge in the future will be identifying benefits for the implementation of area-wide or regional IPM programs. The tactics of regional IPM programs and the benefits they bring to agriculture should be packaged and presented to growers and grower organizations. Regional IPM strategies must be blended with on-farm, site specific tactics in order to complement existing IPM programs. Tighter coordination with state, federal, and private IPM programs is necessary in order to insure efficiency in transferring IPM technology for regional programs.

The grower, private consultant and agriculture related industries have been primary audiences for IPM educational programs. As public visibility and concern over pesticides and dietary risk escalates, innovative methods of educating the consumers and the general public must be developed. IPM has a major role to play in consumer education. Making the public aware of IPM as a reasonable approach to the safe production of food and environmental protection will go far in minimizing panic and misinformation on the safety of food in the United States. Additionally, IPM must be viewed as the appropriate framework in dealing with environmental problems related to pesticides. Educational programs in cotton and other commodities that address environmental concerns through IPM must be addressed.

CONCLUSIONS: A SNAPSHOT OF COTTON IPM IN THE FUTURE

Were we to take a step through time to the year 2012 and drive by any cotton field, at first glance we would notice little if any difference as compared with 1992. Were we to spend a whole season in this future, we would come to recognize that many of our ways of managing and producing cotton had changed drastically. Changes will be at several levels. Some of these changes will be perceived as being restrictive in nature. For example, increasing population pressures and environmental pollution will have brought about an inevitable increase in local, state, and federal regulation of pesticide and fertilizer usage. The majority of changes, however, will be perceived as positive and would have moved agricultural management and production closer to a science.

The most obvious change will be in the amount of conventional pesticides used and how they are used. Conventional application of pesticides will be less than 20% of that used today, with many fields not requiring applications. Much of this decrease in pesticide use will be through the development of integrated production and management systems which rely heavily on the use of resistant and tolerant plant cultivars, and cultural controls that eliminate or severely reduce insect, disease, and weed pests. Such a system will foster an increase in the effectiveness of naturally occurring and commercially released biological control agents. All of this will be achieved with a corresponding increase in yield, quality, and profitability.

By the year 2012, genetic engineering coupled with more traditional breeding techniques will have matured to the point where transgenically altered cotton will confer greater yield, quality, and marketability through insect, disease, and environmental stress tolerance, through basic changes in plant physiology involving carbohydrate and nitrogen partitioning, and through expansion to specialty markets with greater emphasis on colored cottons and on cottons having special fiber characteristics. Resistance to insects and diseases will be broad-based, so as to greatly reduce the chance of resistance developing. Genetically engineered crops will express three or more toxins each differing in their mode of action. The ability of pest species to overcome resistance will be further controlled by the use of phenological and structure specific promoters that control the timing and expression of resistance characteristics. Some of these genes will likely be triggered to produce their toxins by the low dose chemigation of hormonal-like analogs to initiate the expression of the transgenically controlled toxins and antifeedants.

Weed species will be better controlled by the use of cotton cultivars which transgenically produce allelopathic chemicals. Considerable advances will also have been made through the use of biodegradable sprayable mulches. The trend towards narrow row, uniform planting systems will further reduce weed problems while promoting greater crop uniformity. For some types of soils, drill planters will be developed that enable cotton to be directly seeded as part of limited or no-till cropping systems.

With far fewer broad-action conventional pesticides applied, natural enemies will be able to provide adequate control to maintain most outbreak populations of insects, weeds, and diseases below damaging levels. The effectiveness of these biological control agents will be further increased by the use of companion cropping, interplanting, and relay cropping production systems which promote the buildup of natural enemies, and the control of insect, disease, and weed pests. Fields or regions of the cotton belt that have repeated problems with specific pests, will often be addressed by inoculative or inundative release of selected biological control agents. In those instances where a pesticide is warranted, pesticide resistant biological control agents will be released to augment control of insect and disease pests.

By the year 2012, agriculture will see the creation of an entirely new professional position to address the management of cotton as well as other crops. With few exceptions, when growers have an insect, weed, or disease problem that they think requires the use of a pesticide, a Crop Practitioner Doctor will be contacted before an appropriate pesticide can be prescribed. The doctor will have had exhaustive training in the diagnosis and management of agricultural crops, and will be certified in selected areas of expertise. The predominant role of (CES) IPM Specialists will be to train the Crop Practitioner Doctors. CES will also be more fully engaged in field usable research. To determine the appropriate management action to take, the doctor will examine the field, and run a series of standardized diagnostic tests and forecasting management models. The diagnostic tests will involve the use of sophisticated analytical programs which will determine the presence of mineral deficiencies and plant pathogens. The forecasting models will examine, in a matter of minutes, thousands of possible management scenarios, taking into account cost and benefits, including projections of control, impact on non-target organisms, probability of pesticide resistance developing, and the projected environmental fate of each potential management action. After weighing the information provided by the grower, and generated by the certified diagnostic tests and simulation models, the doctor may prescribe the use of one or more pesticides. As part of the prescription, the doctor will provide information on the pesticide concentration, application rate, and mode of application. Local and state regulatory agencies will require that the standardized set of diagnostic exams and forecasting management models be used prior to the use of any pesticides.

As with most medicine, a grower will only be able to obtain most pesticides on a prescription basis. The pesticides will be very different from those used today. When a crop chemical pharmacist fills a prescription, many of the pesticides will be manufactured using highly sophisticated analytical laboratories capable of producing a range of chemicals. The pesticide will usually be highly selective. To reduce the rate of resistance development, the pesticide will also usually differ in its mode of action compared with any pesticides recommended earlier in the season.

By the year 2012, the boll weevil and pink bollworm will have been largely eliminated as pests through areawide suppression and eradication programs. We will see a shift from crop specific field level management to farm and regional level agroecosystem management. This shift will be further fostered by the continuing trend to larger and fewer farming operations. Sophisticated integrated cropping systems management programs will be in common use and will provide regional forecasts of insect development and migration. These programs will serve as aids in planning and designing management production systems, and will enable users to determine which combinations of crops to grow in a region to best consider existing management and production constraints.

These integrated cropping systems management programs will be automatically tied into field level remote sensors and will be used in estimating the buildup of major insect, disease, and weed pests. Crop growth and development forecasts will be routinely conducted, and will be used to examine a range of crop and pest management scenarios, with simultaneous optimization criteria that include economics, soil erosion, and ground water degradation. These models will be used by crop practitioner doctors, farming corporations, regional farm cooperatives, crop production consulting firms, state and federal government regulatory agencies, and by scientific and extension agencies.

Many of the advances predicted to occur over the next 20 years will be developed jointly between government and the private sector. How far we progress towards sustainable agroecosystem management will largely depend on the degree to which we plan for the future, and our success at increasing funding so that true agroecosystem research and extension can be developed.

Cotton Action Team

The following individuals participated in the development of information presented in this report and reviewed the manuscript which resulted from their accumulated thoughts.

Ray Frisbie (Co-Chair)
Texas A&M University

Mike Chandler
Texas A&M University

Harold Coble
North Carolina State University

Jim Devay
University of California, Davis

Bob Frans
University of Arkansas

Andy Jordan
National Cotton Council

D. D. Hardee (Co-Chair)
USDA/ARS
Stoneville, MS

Jim Brazzel
USDA/APHIS
Edinburg, TX

Mark Cochran
University of Arkansas

Kamal El-Zik
Texas A&M University

Johnie Jenkins
USDA/ARS
Mississippi State University

80

Will McCarty
Mississippi State University

Terry Stone
Monsanto Corporatation

Carolyn Theus
Englewood Foundation

Richard Herrett
ICI Americas, Inc.

Tom Kerby
University of California, Davis

Phil Roberts
University of California, Riverside

Ted Wilson
Texas A&M University

Robert Hart
Rodale Research Center

James Supak
Texas A&M University

Chapter 4

BIOLOGICALLY INTENSIVE PEST MANAGEMENT IN THE TREE FRUIT SYSTEM

James P. Tette, Director
New York State IPM Program
NYS Agricultural Experiment Station
Cornell University
Geneva, NY 14456

Barry J. Jacobsen, Professor and Chair
Department of Plant Pathology
Auburn University
Auburn, AL 36849-5409

Formal tree fruit Integrated Pest Management (IPM) efforts began in 1972 when the Huffaker Project, funded by the National Science Foundation, and the Smith-Lever Special Project, funded through the Cooperative Extension Service (CES), provided states an opportunity to develop new knowledge and to transfer knowledge to growers. By this time many states had already developed considerable IPM knowledge for all types of tree fruit and nut crops through regular programmatic efforts, and CES in those states began to transfer that knowledge through formal Extension IPM programs.

Progress in research was marked by interdisciplinary projects which elucidated the relationships and interactions of the knowledge emanating from the various crop protection disciplines. For example, the impact of ground-cover on mites and their predators shed light on how biological control of mites could be influenced by ground cover management. Progress in extension was marked by the preparation of manuals and fact sheets, and by reports of significant reductions in pesticide use as the result of IPM implementation projects.

In the following fifteen years, IPM knowledge was greatly expanded in methods such as pest forecasting, pest monitoring, and the development and use of economic thresholds. This knowledge has allowed fruit growers the opportunity to use pesticides only when needed for most pest problems. While these methods are essential to any IPM program, they often served only to refine the use of chemical pesticides.

In the meantime, pressure from society began to build to further reduce pesticide use as did scientific evidence which suggested the need for alternatives to pesticides. Crop protection scientists and many others recognized that a biologically balanced system was needed to minimize the impact of most pests. Although biologically intensive IPM methods had been studied for several years, it appeared that this research required an excessive amount of time to yield results. In addition, pests were not being totally eliminated with these methods, but were merely reduced in damaging numbers. The lack of private sector involvement in the research effort, and the lack of vision from federal funding agencies, meant that the states had to bear most of the costs for development of biologically intensive IPM methods.

Many different factors spurred a change in direction away from pesticides and toward increased attention to IPM and biological management of tree fruit. The single greatest mobilizing influence was probably the daminozide controversy. Now it is clearly recognized that conventional pest control is at risk, that tree fruit production must become a more sustainable system, and that environmental stewardship will continue to be an increasingly important consideration.

CONVENTIONAL PEST CONTROL APPROACHES APPEAR TO BE "AT-RISK"

Today, pest resistance to pesticides is increasingly common and with fewer pesticide choices this problem will only intensify. The pesticide reregistration process is substantially reducing the number of chemical alternatives that can be used. Thus the strategy of altering pesticide types to reduce the occurrence of resistance will be limited. This loss is acute for all minor crops such as tree fruit. The primary reason for these losses is the expense to the registrant for reregistration. Food safety concerns and the use of pesticides are perceived by many as incompatible. There is a general perception that natural controls (e.g., those which are biologically-based) are safer than synthetic pesticides. The risk to producers of a single pesticide residue being detected which could be perceived as harmful (e.g., Daminozide in apples) is very high, and since relatively few pesticides will be available in the future it is more likely that a single pesticide residue would result in such a reaction. Finally, farm worker safety as affected by pesticide use is a problem of increasing importance. This is a serious problem for labor intensive crops such as tree fruit.

Agriculture Needs to Move Towards More Sustainable Systems

Sustainable pest control systems require a diversity of pest control tools or tactics. Biological controls and other biologically intensive methods will be an important part of sustainable pest control systems. Biologically intensive IPM systems will be responsive to changes in environment, cropping systems and new pests, and will be competitive in changing economic climates. Biologically intensive IPM methods will help address both groundwater and nonpoint source pesticide pollution. Such methods will also be an alternatives where urbanization and wildlife concerns interface with production.

BIOLOGICALLY INTENSIVE METHODS OF PEST MANAGEMENT

There are many ways to categorize biologically intensive IPM methods. We choose to break these methods down into four areas: biological controls, cultural controls, host plant resistance, and semiochemicals. Further delineation would include biological controls using macrobes (parasites, predators, and other natural enemies) and microbes (viruses, fungi, bacteria, protozoa, entomogenous nematodes); host plant resistance using conventional breeding methods or transgenic gene insertion; cultural controls using ground cover management to remove inoculum and to enhance the establishment and efficacy of biocontrol agents, cover crops, mulches, pruning, tree structure, rotation, and removal of wild hosts; semiochemicals using pheromones, kairomones and allomones.

Efforts aimed at the commercialization and adoption of these methods have experienced significant constraints from technical, regulatory, economic, and institutional barriers. These barriers must be overcome if biologically intensive methods are to become a common part of commercial agriculture. In constructing this report, we have tried to identify these constraints as specifically as possible. Solutions are expressed in the form of a plan. No attempt was made to provide an in-depth look at the state of any one method since not all fruit growing regions were able to respond to the survey.

In constructing this report we have tried to incorporate both constraints and solutions. The constraints or barriers are expressed as specifically as possible, while the solutions are expressed in a proposed action plan. We do not attempt to provide an all inclusive assessment of the state of the art of specific methodology since not all fruit growing regions responded to our questionnaire.

THE NEED FOR FUTURE RESEARCH ON BIOLOGICALLY INTENSIVE IPM METHODS

Biological Control

Most biological control research projects are designed to study the management of pests through the influence of predators and parasites. Other studies involve the use of augmentative releases to enhance existing populations of beneficials. A few examples of genetic improvement of beneficial arthropods were reported through selection of insecticide resistant predatory mites. Research was reported on biocontrol of mites in six states and on biocontrol of insects in four states. Grower adoption of biological control methods using predatory mites to combat harmful mites was reported in 13 states. Reports from six states indicated various forms of biological control of insects with naturally occurring predators and parasites were being utilized by growers.

Microbial forms of biological control are primarily delivered as an application to a crop or plant as needed. Research on the use of microbials for insect control by a virus was reported in two states, by bacteria in three states, and by entomogenous nematodes in one state. Research using biocontrol with naturally occurring microbial agents to combat harmful pathogens was reported in four states. Similarly, postharvest pathogen management using foliar methods was reported in one state. In our survey only one state reported the use of an insect virus by growers, while most states indicated growers were using bacteria against insects. Reports from three states indicated growers are using biological control agents to combat pathogens.

Host Plant Resistance

Conventional breeding programs for disease resistant cultivars were reported in three states for apples and in one state for peaches. Breeding for resistance using transgenic gene insertion methods was reported in two states. Reports from nine states indicated grower use of disease resistant cultivars of apples, and one state reported the use of similarly resistant cultivars of peach.

Cultural Controls

Research on the use of cultural controls was reported in nine states. Reports from sixteen states indicated growers are using some type of cultural control methods.

Semiochemicals

Research on the use of pheromones to disrupt insect communication associated with reproduction was reported for a number of pests including obliquebanded leafroller, navel orangeworm, various leafminers, peach twig borer and tufted apple bud moth in nine states. Research on plant produced volatiles which are used by insects is also being conducted for several species. Two states reported mating disruption with pheromones was being practiced by growers of apples, peaches and almonds for codling moth, peach tree borer and oriental fruit moth. One state indicated that growers were using semiochemicals combined with sticky traps to mass trap apple maggot. Another state reported the use of sterile male releases to combat insects.

FUTURE RESEARCH ON BIOLOGICALLY INTENSIVE IPM METHODS

The highest priority expressed by researchers responding to our survey was understanding the basic biology and ecology of pests and biological control agents to continue the development of biologically intensive IPM methods. This was closely followed by the need to focus on the application of biological control in commercial orchards. Specific biological control projects included isolation of pathogens and other microbial agents which would aid in the management of summer diseases, exploration for parasites and predators of leafhoppers, aphids, leafminers and leafrollers, and the use of parasitic nematodes for tree fruit insects which overwinter in the soil.

Continued development of techniques to disrupt mating of foliage feeding insects was also highly rated as many of these insects are in fact secondary pests, and pesticides applied for their control often disrupt other biological control systems. Research on the ecology and biology of the phyllosphere, or leaf fruit surfaces, could lead to the development of non-toxic foliar modifying agents which would prevent pathogen attack.

Host plant resistance research could be enhanced through the identification of resistance mechanisms. Because long time periods are required for cultivar development, new biotechnology techniques such as tissue culture screening and genetic modification should be supported. There is a critical need to develop fruit cultivars resistant to late season pathogen attack.

Because biologically intensive IPM may require changes in fruit cosmetic quality standards, it is important that socioeconomic research be conducted with regard to consumer acceptance, market value, etc. Large scale, long-term research on the effect of alternative pest control methods or new production systems on farm profitability must be conducted if producers are to adopt new technology.

Cultural controls such as altered orchard configuration, ground cover management, irrigation, etc., need to be thoroughly examined. It is important that such systems be evaluated not only for their potential use as biologically intensive IPM techniques, but for productivity and profitability as well.

THE ROLE OF COOPERATIVE EXTENSION

Most respondents indicated that the role of Cooperative Extension (CE) will continue to change. There will be increased demands for Extension to play a key role in the development and transfer of biologically intensive IPM methods. This involvement is viewed as critical to the successful implementation of biologically intensive methods. More than any other organization, CE has the best opportunity to organize interdisciplinary teams to demonstrate concepts and to address problems. Adaptive research to adjust biologically intensive methods for different regions or localities can best be carried out by CE. Extension specialists should also be able to measure the impact of the application of biologically intensive methods to provide the data to guide the further development and use of these methods. Working with private and public organizations, CE should conduct large scale demonstrations of biologically intensive IPM methods over five to seven year periods. These projects should be organized to insure that economic and environmental impacts would be measured. CE needs to involve consultants more directly in the planning and execution of demonstration projects and provide more direct access to program results. Their educational role should not be confined to growers, practitioners, and pest control advisors, but must be expanded to consumers and environmental groups. Extension specialists need to re-examine the structure of their workshops, conferences, newsletters, and publications to facilitate the adoption of biologically intensive methods. Their crop protection literature should begin to reflect alternatives to pesticides, and include ways to reduce or avoid pesticide use. Pesticide Applicator Training programs conducted by CE should be restructured to include or even emphasize biologically intensive IPM methods.

CONSTRAINTS TO THE USE OF BIOLOGICALLY INTENSIVE METHODS IN TREE FRUIT

Federal and State Regulations

Some federal and state regulatory policies severely constrain the development and use of biologically intensive IPM methods. Although technical and commercial impacts are often mentioned, these constraints impact the economics and adoption of biologically intensive IPM methods as well. Barriers and constraints identified by the Tree Fruit Action Team, in order of priority include:

High Priority
   a. Unreasonable restrictions on testing procedures and plot size which slows research on new technology, and delays development and adoption.
   b. Lack of clear criteria and pathways for registration, testing, etc.
   c. Application of "older pesticide-type" registration mentality and criteria to the registration process for "new" pesticide chemistry.
   d. Unreasonable crop destruction requirements. For products not applied to crops researchers often must follow the same crop destruction requirements as those of conventional pesticides. This runs counter to the need for large scale testing of these materials.

Medium Priority
   e. Poor interpretation of efficacy standards. The interpretation of efficacy standards appears to be too severe for nontoxic pest control method candidates.
   f. Regulation of genetically modified organisms inhibits their intrastate exchange and testing.
   g. Regulations on the use of a patented clone as a parent in a breeding program are unduly restrictive.

h. Registration requirements which call for separate registration packages for each strain or strain combinations. Current regulations may prohibit the free exchange of microbial germplasm for use in improvement of local strains of microbial biocontrol agents.

i. Conflicting state and federal requirements. Some states have agricultural codes which prevent the sale of products having a federal Experimental Use Permit (EUP) unless a conditional registration has been issued by the state. Producers of natural enemies find their businesses hampered by regulations that differ from state to state. Some states require that regular samples of natural enemies be sent to their regulatory agencies. The registration processes of individual states may preclude the use and large scale testing of promising biologically intensive materials for many years.

j. Lack of distinction in the regulatory process. The same EPA product managers review safety data for both toxic and non-toxic candidate products.

k. Economics of the registration process. The registration process was judged to be too costly and time consuming for most biologically intensive agents.

Lower Priority

l. Commodity grading standards. Current fruit quality standards may be difficult for growers to meet using biologically intensive IPM practices or using production practices that limit pesticide use. New grade standards should be considered that reflect the true quality not just the cosmetic quality of fruit.

m. High nonspecific standards for insect parts in processed foods. Many of the Food and Drug Administration (FDA) requirements for insect parts in processed food preclude the use of biocontrol agents.

n. Lack of access to key people in regulatory agencies. As registration processes are pursued it is important that key regulatory people be more accessible for consultation and advice.

o. Lack of accelerated process for biologically intensive products. Many fruit workers feel there is a slow process for approval of EUPs.

Nonregulatory Constraints

Numerous nonregulatory barriers and constraints were identified by the Tree Fruit Action Team including:

High Priority

a. Lack of incentives to manufacturers, developers, and farmers. It is often difficult to gain proprietary control of biological control organisms, or to encourage growers to use biologically intensive methods.

b. Lack of resources directed toward biologically intensive IPM. Funding is lacking for interdisciplinary research, both basic and applied, in the Land Grant University system and for cooperation with small companies developing biologically intensive products. Funding is also lacking for biologically intensive research on minor crops (cherry, apricot, peaches). At present there is a lack of focus on biologically intensive IPM, due primarily to the higher risk associated with these research, development and education efforts. Such work requires long-term support that is not presently available. Resources directed to "chemical systems" commonly exceed $40 million for the registration of a single pesticide. This level of funding is not available for all biological control research nationally.

c. Economic constraints. For most growers it is cheaper to apply pesticides to high value crops than to apply biologically intensive methods. Biologically intensive methods are often more variable in their performance than pesticides and their availability may be more limited. Many companies have only small production facilities for the manufacture of biologically intensive methods. Further, the companies may have a limited financial base, and therefore have more limited time to develop a product with payback potential. Similarly, providers of natural enemies usually do not have the capital to invest in the necessary rearing facilities. Private consultants need to acquire more information and ensure greater management where biologicals are used. There is also a greater risk of failure (hence loss of employment), therefore consultants must charge more for implementation of biologically intensive IPM programs.

d. Institutional constraints especially in the Land Grant system where scientists must "publish or perish." This syndrome usually leads to a prevalence of short term research which is often reinforced by federal funding programs.

Medium Priority
    e. Lack of coordination of federal and state agencies.
    f. Lack of trained and certified consultants. Everyone agrees that biologically intensive methods will be more difficult to implement and will require the expertise of private consultants, or an increased role for CE with increased resources.
    g. Societal constraints. Demands by retail grocers and the public for blemish free fruit impede the development and use of biologically intensive IPM methods. Many retailers demand cultivars which they can get in large quantity over an extended portion of the year. Then too, the increasing public concern over genetically engineered crops will constrain adoption. An associated constraint is public perception that private industry should bear the cost for registration of biologically intensive methods.
    h. Knowledge constraints. The lack of environmental impact assessment information and documentation of these impacts impedes progress on biologically intensive methods. For example, with concepts like natural enemies of pests there is a lack of information on life cycles of many natural enemy species, their environmental tolerances, feeding and reproduction rates, and a lack of guidelines on how to apply them. There is also a lack of information on the relative compatibility with different pesticides, and quality control constraints associated with mass production, storage and shipping. These methods also lack a sales and support system comparable to that of the pesticide industry. There is a lack of knowledge on how to educate consumers. There is a lack of knowledge for growers and consultants on overall management considerations, how to deal with lower efficacy, and what to expect in terms of risks/benefits from a biologically intensive IPM program.
    i. Technical constraints to biopesticides often include: poor residual activity, narrow host range which limits market potential, inconsistent control associated with longer kill time or limited virulence, high production costs, lack of technology for mass production, and formulation and delivery problems.

Lower Priority
    j. Constraints associated with pesticide resistance. Pesticide resistance encourages frequent use of high concentrations of pesticides which in turn destroy natural enemy complexes.

SUGGESTIONS FOR OVERCOMING CONSTRAINTS TO THE DEVELOPMENT AND USE OF BIOLOGICALLY INTENSIVE IPM METHODS

The Tree Fruit Action team and the survey respondents had several suggestions for both short and long-term options to help remove the contraints facing biologically intensive IPM. These suggestions could form the basis of a plan to influence policies and define roles for all partners. This plan will require the cooperation of scientific, educational, regulatory and private groups and organizations. The suggestions are as follows:

The Environmental Protection Agency

- Allow large scale testing of biologically intensive materials without the requirement for crop destruction. The acreage to be included in the testing program should depend on the mode of action of the technology.

- Develop a new system of issuing permits for research without restricted plot size and crop destruction requirements. We would suggest they be called Experimental Research Permits (ERP) and that they be issued only to Land Grant or USDA scientists. These permits would allow early (in the development stage) evaluation of a product. Some restrictions could be placed on acreage and proximity to water sources.

- Adjust regulatory policies to reduce requirements and accommodate the registration of biologically intensive methods. Provide a clear regulatory pathway for biologically intensive methods that is different from synthetic organic pesticides. Consider the nature of the pest control agent when determining the safety tests needed and provide a separate fast agency track for processing. Use generic registrations for certain viruses, pheromones, biologicals, etc. Change regulatory procedures so that effects of a regulatory action on a given pesticide considers the resulting efficiency of remaining pesticides and beneficials.

- Assign product managers to biologically intensive methods only.

- Prepare label requirements in conjunction with Land Grant input by establishing a Land Grant/EPA liaison team to assist in the assessment of candidate biological control agents.

- Develop a plant health prescription program which would govern the application and use of biologically intensive, chemical, and other crop protection methods.

- Develop a plant health prescription policy to dissociate chemical sales from the recommendations of materials.

- Support the development of a cadre of independent IPM practitioners and encourage growers to utilize their expertise. Independent practitioners will use the best control technology, not one that will produce income from pesticide sales.

- Provide a document which clearly outlines the regulatory process for companies, Land Grant personnel, and others working on the development of biological agents.

- Relax rules for interstate movement of nonpathogens and/or their introduction.

- Establish uniform federal laws which supersede state laws.

- Accept toxicological and other pertinent data on products from foreign sources if they meet EPA guidelines, and good laboratory procedures.

State Regulatory Agencies

- Follow EPA guidelines for registration requirements, and avoid imposing unique regulations which would inhibit the testing or use of biologically intensive methods unless it can be demonstrated that there are special local circumstances justifying change.

- Enforce a national plant health prescription policy.

The Food and Drug Administration

- Develop a policy which would allow more defects or blemishes on fresh fruit. Revise the acceptable level of arthropod parts in processed food so that levels are set according to acceptable appearance of the final product and hazards to human health instead of the current arbitrary standards.

The United States Department of Agriculture

- Provide funding from a national perspective by rewarding and stimulating crop-oriented regional centers of excellence for research in various types of biologically intensive methods.

- Stimulate the creation of similar centers of excellence for the evaluation and implementation of biologically intensive methods.

- Direct new incentive programs like the USDA Agricultural Stabilization and Conservation Service Integrated Crop Management effort toward use of biologically intensive IPM methods.

- Provide a crop insurance program and an IPM certification program for growers who use biologically intensive methods and who practice IPM.

- Provide an incentive program to induce fruit growers to utilize pest resistant cultivars.

- Maintain strict accountability and reporting for the use of funds for research and extension efforts.

Animal and Plant Health Inspection Service

- Relax rules for interstate movement and introduction of predators and parasites.

- Relax or streamline rules on interstate movement and testing of both genetically modified and unmodified biologicals. These regulations are excessive for many potential biocontrol agents. Certain genetic markers should be given blanket approval.

95

- Provide funding to find more effective natural enemies, and to develop advanced production systems for commercial insectaries.

- Permit large scale field trials to determine the practicality of using natural enemies.

The Land Grant System

- Create interdisciplinary centers of excellence for biologically intensive research, or centers for evaluation and implementation of biologically intensive methods. Remove the barriers to interdisciplinary research and implementation.

- Develop methods which simplify and synthesize IPM techniques for producers or consultants. For example, "expert" computer consultant packages could be developed to aid in IPM decision-making.

- Through CE and working with the Soil Conservation Service (SCS), conduct long-term demonstration programs of biologically intensive systems.

- Provide education for growers and consultants in the use of new methods.

- Educate growers, distributors, consultants, and others on techniques which minimize pesticide resistance.

- Develop a consumer education program in conjunction with retailers and processors. Provide leadership in educating the general public in accepting new cosmetic standards, cultivars and products.

- Monitor resistance levels in pest populations on a systematic and continuing basis.

Private Companies

- Develop an economic incentive marketing program when first introducing a biologically intensive product to stimulate its adoption in the first three years of commercialization.

- Develop a policy of product introduction which includes close coordination with the Land Grant system.

Retailers and Food Processors

- Offer growers and producers a premium price for fruit products grown with fewer conventional pesticides. For example in the East it could be possible to offer a premium price for disease resistant apples.

- Educate consumers and promote their purchase of fruit grown using biologically intensive methods. Educate consumers about the benefits of IPM.

- Develop markets for IPM produce, through programs which display fruit certified as grown using IPM. "IPM Certified" must have standards developed in accordance with a coordinated plan including representation from EPA, USDA and the Land Grant system.

Private Crop Consultants

- Utilize and evaluate the IPM methods and techniques developed through research rather than relying strictly on experience.

General Suggestions

- Strive for a balanced approach to IPM by maintaining diversity in possible pest control tactics, including pesticides with different modes of action.

- Have all federal agencies whose efforts and policies impact crop protection identify funding for a national, large scale IPM programmatic effort. Make much of this funding available as long-term block funding to speed the development of biologically intensive IPM methods.

- Modify United States patent laws, where appropriate, to encourage the commercial development of biological control agents.

- Foster biologically intensive methods developed for small markets through special, low-cost loans and technical assistance to small companies.

- Determine the economics of pest control for both traditional and biologically intensive IPM methods. Where possible, include cost of externalities such as groundwater quality, food safety and farmworker safety.

- Minimize risks to producers, registrants and consultants through crop loss insurance, price guarantees, or government/grower cost sharing for scouting.

- Ensure minimum use of pesticides with high risk of pest resistance, and utilize tactics which would reduce the likelihood of resistance.

- Utilize the existing structures of the IR-4, NAPIAP, and BAP programs to minimize resistance, and optimize the potential for biologically-based crop protection products.

The adoption of biologically intensive IPM methods in tree fruit will permit the development of a less toxic approach as a solution for pest management. As a result, pest pressures on tree fruit crops will shift or change through natural processes of evolution. These shifts will necessitate new learning and continued development of new strategies.

If biologically intensive IPM methods are to be developed, two ingredients are necessary. A framework is needed to incorporate the numerous ideas and concepts that exist, and a goal or ideal is necessary to indicate the direction that needs to be taken. This "blueprint for the future" is the subject of the remaining text.

BLUEPRINT FOR THE FUTURE

According to USDA statistics, tree fruits or nuts are produced in almost every state in the nation. For purposes of simplicity one cropping system, apples, was chosen for detailed analysis, and for the preparation of a blueprint for the future. Commercial apple production takes place in 36 states ranging from a low of 5.5 million pounds in Rhode Island to a high of 5 billion pounds in the state of Washington. Pest problems are numerous and somewhat defined by the region in which the crop is grown. This blueprint does not attempt to identify specific regional differences, it merely acknowledges the complexity of pest management in this cropping system.

A Conceptual Model of an Apple Orchard Without Pesticides

In developing a blueprint for an apple orchard of the future, the working group was asked to assume pesticides were not available. This assumption carries with it the understanding that significant amounts of new funding will be available for both research and technology transfer efforts. Without such funding, the objectives of this design will either take decades to achieve or may never be achievable. The optimal design is based upon pest management decisions that must be made by the grower.

Site Selection and Preparation
The site will be selected to accommodate desired environmental conditions for proper air drainage and rapid drying of foliage. Nematode management will be achieved by preplanting the site to marigolds or other nematode suppressive plants the year before the trees are scheduled to be planted.

Tree Selection
The rootstock for the tree will be size controlling, certified virus free, and resistant to voles and fireblight. The cultivar will be selected for resistance to various fungal pathogens.

Cultural Practices
The orchard floor will be planted to slow growing and/or genetically dwarfed grass cultivars. Propane burners will be used to control weeds and vegetation in the tree row, and will also provide control of voles and some insects.

Cultural practices such as sanitation, Y-trellising, summer pruning, and fruit thinning will be implemented. Most of these practices are labor intensive and will require an adequate available labor pool and an efficient labor management program.

In-Season Disease, Mite, and Insect Management
Disease management not previously addressed will be approached with biological control agents applied to leaf and fruit surfaces using electrostatic sprayers. Insects and mites will be managed with natural and introduced biological control agents, microbial pesticides, pheromones, horticultural oils, repellants and traps.

Postharvest

Postharvest pest problems associated with the fruit, mostly pathogens, will be managed by dipping the fruit in water baths containing biocontrol agents. Controlled atmosphere storage will also be used for disease management as appropriate. Pest problems associated with the tree, mostly damage by voles, will be managed with tree guards and repellents.

Grower Awareness

Technology transfer will be planned and delivered by a combined team of CE agents, SCS personnel, certified IPM private consultants and growers. This transfer process will include:

1. Long-term demonstration on key grower farms with indemnification for crop loss.
2. Economic analysis of these demonstrations.
3. The use of a computerized decision-making and record-keeping system which will "coach" growers in proper pest management decision-making, and provide a record of their management decisions.
4. A public education component to continually explain the results of the IPM effort, and to encourage the consumption of products from these farms.
5. Fruit will be marketed to environmentally conscious consumers who will accept no blemished fruit, and who are willing to pay a slightly higher price for their apples.

Other Key Elements

One key element which must be established in order to achieve the optimal design is that biologically intensive products must be commercially available to growers. Products are of no value if discovered, but still residing in the cupboards (or journals) of the research scientists.

Timetable for Adoption of the Design

Research on each element in this design is now underway at Land Grant Universities in the United States. In addition, some biologically intensive products are currently available from private industry. However, this design will not be attainable in the next 10-20 years unless there is a major effort to remove federal and state constraints to the testing and validation of biologically intensive methods. Major infusions of funding will also be required for the research and implementation efforts.

Reliance Upon Multiple Methods

The dynamic nature of a tree fruit system means that the conceptual model will constantly change through time. In this respect it becomes necessary to modify the blueprint in order to arrive at a biologically-balanced pest management system for apples by returning available chemical pesticide methods to the optimal design. This will allow the system to rely upon many forms of pest management, none of which will require a "silver bullet" solution to pest problems which may prove difficult to attain. Chemical pesticide methods which are least toxic to all elements of the environment (applicators, farmers, consumers, natural control organisms, ground and surface water, air quality and wildlife) would be added to the design. In this blueprint, recommendations for the use of these materials will be made on a prescription basis which will require the submission of data supporting the need for such use.

Research Progress Towards Achieving the Optimal Design

A detailed analysis of current research on biologically intensive methods was developed through a survey of fruitworkers nationwide. A synopsis of this research is provided in Table 1. The actual extent of research now underway is thought to be much greater as not all fruitworkers responded to our survey.

Two trends were apparent. Most of the ideas expressed in the conceptual model of this apple orchard are currently being researched in the Land Grant University system, and many are nearing the point of application. The major stumbling block continues to be the lack of funding for the applied research and development aspects of the effort, as well as for more fundamental research efforts. Funds to demonstrate these concepts, which are primarily provided through CE, are similarly not adequate.

Table 1. The availability and status of research on biologically intensive methods in the apple system.

Pest	Biologically intensive methods currently available	Research underway	Estimated time frame before implementation	Other research leads
Aphids Apple, Rosy, Wooly, etc.	None	Biocontrol using endemic predators and parasites	3-5 years	Alarm pheromones
Apple Maggot	None	Trap out using superattractive poisoned spheres Semiochemical ovipositional deterrents	20 years	Breeding apple maggot resistant apples
Coding Moth	Biopesticide (Bts) Granulosis virus	Pheromone disruptants	Commercial tests underway now	None reported in survey
Mites European Red Two spotted	Biocontrol using indigenous predators	Introduction of "foreign" predatory mites Development of pesticide-resistant strains of predators.	3-5 years	Resurgence of natural control in a nonpesticide environment
Green Fruitworm	Biopesticide (Bts)	None reported in survey	NA	None reported in survey
Leafrollers	Biopesticides (Bts)	Mating disruption Pesticide combinations using 1/10th rates of pyrethroids and full rate of Bt	1-2 year	None reported in survey
Plum Curculio	None	None reported in survey	NA	Biocontrol with entomophagous fungi
Mirids Tarnished Plant Bug	None	Biocontrol with introduced predators	3-5 years	None reported in survey

102

Table 1. (continued)

Pest	Biologically intensive methods currently available	Research underway	Estimated time frame before implementation	Other research leads
Leafminers	None	Biopesticide (IGR) diflurbenzuron needs registration	1 year	Understanding of parasite dynamics and releases
White Apple Leafhopper	None	None reported in survey	NA	None reported in survey
Apple Scab	Resistant cultivars are available	Evaluation of biocontrol agents	5-10 years	Naturally occurring biocontrol agents
Powdery Mildew	Resistant cultivars are available	Evaluation of biocontrol agents	5-10 years	Naturally occurring biocontrol agents
Summer Diseases	Cultural and sanitation measures	Better understanding of infections Bleaching-out of symptoms on apples	3-5 years	Biofungicides
Postharvest Decays	Cultural sanitation measures	Biocontrol using yeast and/or bacteria	5-10 years	Use of oxygen in controlled atmosphere storage
Weeds	Flaming	Using mulches and cover crops, or mechanical cultivations for ground cover management	3-5 years	Dwarf grasses
Whitetailed Deer	Fences	The economics of using fences	3-5 years	Planting a diversionary crop Determine sex ratio of browsing herds by spotlight counts
Voles	Ground cover management, tree guards	Resistant rootstocks Relationship between ground cover composition and vole activity	5-10 years	Adding raptor perches and nest boxes in orchards
Woodchucks	None	None reported in survey	NA	None reported in survey

103

CONCLUSIONS

While biologically-balanced tree fruit production systems are the desired goal, many of the intended adopters (fruit growers) remain skeptical as to the value of biologically intensive methods. Many growers have already made thoughtful efforts to incorporate already existing methods, but many have experienced financial losses because these new methods may not provide the desired results immediately. Biologically intensive IPM methods will require long-term research support to apply fundamental discoveries to the field. They will also require even longer periods for adaptation, evaluation and demonstration before fruit growers will rely upon them as they now rely upon chemical methods.

PREPARATION OF THIS REPORT

This report was drawn from forty-two responses to a survey sent to tree fruit workers located throughout the United States, from a meeting of the Tree Fruit Action Team in Orlando, Florida, February 23, 1991, and from a second review of this document by all Action Team members. Respondents to the initial survey included individuals with extensive knowledge of tree fruit production and postharvest management in the private sector, the Land Grant university system, and federal agencies. Survey respondents provided information on apples, pears, peaches, almonds, walnuts and pecans. The "Blueprint for the Future" was developed in conjunction with team members at the Wingspread Conference which was held on August 27-29, 1991, in Racine, Wisconsin.

Tree Fruit USDA/EPA Team

The following individuals participated in the development of information presented in this report, and reviewed the manuscript which resulted from their accumulated thoughts:

Jay F. Brunner
Department of Entomology
Washington State University
Wenatchee, WA

Harvey Reissig
Department of Entomology
Cornell University

Vince Morton
Ciba Geigy Corporation
Greensboro, NC

Arthur Kelly
Fruit Grower
Acton, ME

Hugh W. Ewart
Northwest Horticultural Council
Yakima, WA

Louis A. Falcon
Department of Entomology
University of California
Berkeley, CA

Kenneth D. Hickey
Department of Plant Pathology
Pennsylvania State University

Barry Jacobsen, Co-Chair
Department of Plant Pathology
Auburn University

George Norton
Department of Agricultural Economics
VPI and SU
Blacksburg, VA

Teryl Roper
Department of Horticulture
University of Wisconsin

Terry Schettini
Rodale Research Center
Kutztown, PA

James P. Tette, Co-Chair
NY State IPM Program,
Cornell University

Patrick Weddle
Weddle, Hansen & Assoc.
Placerville, CA

Chapter 5

BIOLOGICALLY INTENSIVE IPM FOR VEGETABLE CROPS

Frank G. Zalom, Director
Statewide Integrated Pest Management Project
University of California
Davis, CA 95616-8621

William E. Fry, Professor and Chair
Department of Plant Pathology
Cornell University
Ithaca, NY 14853-5908

Vegetable production in the United States exceeds $7.5 billion annually. Virtually all Americans eat vegetables daily, either as fresh or processed products. Perhaps more than any other agricultural product group save fruit, consumers have come to expect an assortment of vegetable products for extended seasons, and the availability of fresh vegetables in particular has increased dramatically in the last 30 years. Consumers also expect that the vegetables they eat are of high quality, are unblemished and are relatively inexpensive. Growers have responded by producing vegetables for extended seasons in a variety of different production areas where such production is possible. Packers and processors have met consumers' demand by building an infrastructure to distribute the products demanded. This value added to products increases the total value of the vegetable industry dramatically.

Consumer demands for a dependable supply of high quality, unblemished vegetable products at relatively low cost has made the control of pests which attack vegetable crops an essential part of vegetable production. When left unmanaged, pests including weeds, disease-causing organisms, insects, mites, nematodes and vertebrates have the potential of severely reducing both quality and yield of all manner of vegetables produced.

Pesticide use on vegetable crops became widely used after World War II as pesticides became relatively easy to obtain, and were cost effective. Their effect upon pest populations was immediate and often dramatic. The amount of pesticide used on all crops in the United States increased 19,000 fold from 1940 to 1976 (22). Pesticide use has stabilized since that time, and the amount of pesticides applied has actually decreased about 14% overall since 1982 (19). In contrast, pesticide use on certain crops and pests has increased.

Vegetable growers in particular have profited from the availability of synthetic pesticides. They gained the ability to produce some crops in otherwise unsuitable locations, taking advantage of valuable market niches. Their crops could be grown in uniform plantings on larger than traditional acreages. Seasons were extended to satisfy market demand or to supply local packing or processing industries. Pesticides became widely used after harvest to maintain the quality and to extend the shelf life of agricultural products. As a result, some of the most important traditional pest control methods such as crop free periods, rotations, cultivation and various types of sanitation practices were used with less frequency due largely to economic, land use and labor considerations. In the absence of action thresholds and nonchemical alternatives, growers use pesticides at least in part to reduce production risks. In addition to their value in reducing risk, pesticides are perceived by many producers as being relatively easy to use and less time consuming than other approaches. They also help to reduce variability in production that might be expected with differing pest pressure between years and growing regions.

Recently, the importance of vegetables as consumer noticeable[a] products has been confirmed as the public focused attention on food safety issues. Some food stores are using the issue as a marketing vehicle certifying that their vegetables contain "no detectable residues."

PROBLEMS RELATED TO PESTICIDE USE

There are over 60 different types of vegetable crops grown commercially. Pesticides used on these crops must be registered for the individual crop. In some states such as California, where over 55

---

[a] Food items which consumers purchase in the marketplace and have relatively little value added.

vegetable crops are produced, separate state registration is also required. Most vegetable crops are produced on limited acreage relative to program crops of national prominence such as cotton and corn. As a result, the economics of pesticide registration would dictate that most pesticides address the major acreage crops and that relatively fewer pesticides should be available for use on specific vegetables, and this is the case. The limited number of products registered for use on most vegetable crops makes them especially vulnerable to the loss of specific pesticides. It is estimated that the Federal Insecticide, Fungicide and Rodenticide Act (FIFRA) reregistration process alone will affect over 4,000 pesticide/crop combinations. The majority of these losses affect vegetable crops.

Although there has been much attention paid to recently enacted or proposed legislation that would affect the availability of certain pesticides or crop uses of pesticides, other factors have been responsible for limiting their availability to growers. These factors include pest biology, effect on nontarget species and human health issues. Vegetable crops are especially vulnerable to loss of pesticides because of the relatively few pesticides registered on limited acreage "specialty" crops, the volume per acre of pesticides applied, the number of applications per season, production in areas of specialized climate and soils and intensive labor requirements.

Pest Resistance

Pest resistance is a problem whenever management of a pest species is dependent upon one tactic. Of particular interest on vegetable crops has been the development of pesticide resistance in such key pest species as diamondback moth, Colorado potato beetle, tomato pinworm, several species of aphids and the lettuce downy mildew fungus. When resistance develops in such key pests, production of the host crops can actually be limited or eliminated in a given region.

Cross-resistance of a pest to different classes of pesticides with similar modes of action are also documented. Recently, resistance in insects to pyrethroid insecticides has been found to be widespread and to occur relatively quickly when the target species had developed high levels of resistance to organochlorine insecticides. Of course, organochlorines such as DDT had been widely used in many vegetable growing areas in the past.

Even biological alternatives are not immune to the phenomenon of pest resistance if they are the sole means of controlling a pest species. It is interesting to note that pesticide resistance has been induced in laboratory colonies of tobacco budworms exposed to tomato plants containing genes expressing the toxin of *Bacillus thuringiensis*, a widely used microbial pesticide. Recently, resistance to *B. thuringiensis* has also been identified in Hawaiian diamondback moth populations where the pesticide has been used extensively for control on vegetable crops.

Due to selection against the resistance traits, it is occasionally possible to again use a pesticide effectively for brief periods after a break in its application for an extended period. Usually, however, the effectiveness of the pesticide in such cases is limited to no more than a few applications before the pest populations again show a high level of resistance. There are no proven programs for managing resistance in populations once resistance is identified. The best strategy for managing resistance is to not rely on the application of pesticides as the sole management strategy, and to alternate the classes of pesticides or approaches used so that the pest population is not under severe selection pressure from a particular material. For example, the combination of recently registered narrow spectrum fungicides with broad-spectrum materials would be an excellent management approach and would likely reduce pressure to the development of resistance. Fungicide resistance will likely increase with the loss of broad-spectrum fungicides such as the EDBCs.

Secondary Pest Outbreaks

The emergence of secondary pests has been especially well documented on several insect and mite pests of vegetables. Leafminers, some species of aphids and whiteflies, and spider mites are all considered to be pests of vegetables that are usually under good biological control in the absence of pesticides. When insects such as leafminers and whiteflies reach damaging population levels, additional treatments are required for their control. These pests often have a high level of resistance to the pesticides applied because they have been exposed to the same pesticides as the target pests. Therefore, when they reach damaging levels they are often very difficult to control, and can become limiting to production.

Human Exposure

Farm worker exposure to pesticides is always a potential hazard of vegetable production as most vegetables are labor intensive during a relatively short season, and farm workers are often in fields performing various tasks. Pesticide label guidelines regarding reentry into treated fields and protective equipment to be worn are intended to protect agricultural workers. Because vegetables are

110

typically produced in warm conditions, safety equipment is especially uncomfortable to wear. More training is needed particularly for those workers whose primary language is not English. Labelling in Spanish or other relevant languages would also be helpful to assist workers or producers whose primary language is not English.

Food safety has become an issue of concern to consumers recently even though testing programs have revealed that virtually all harvested vegetables have levels of pesticide residues within Environmental Protection Agency (EPA) established crop tolerances. Evaluations of pesticide residues remaining on fresh and processed vegetable fractions and the establishment of harvest intervals are required as part of the pesticide registration process, and are intended to protect consumers.

CURRENT STATUS OF PEST CONTROL IN VEGETABLE CROPS

It is generally accepted that pesticides are the most widely publicized tactic for controlling many pests of vegetable crops. However for some pests, production areas and crop uses, pesticides are used sparingly or not at all. This makes it difficult to provide a general overview of pest management in vegetable crops that would be representative of all situations. For example, in Arizona and California, generally dry conditions exist during much of the growing season reducing the amount of fungicides necessary for control in comparison to the eastern and central United States where rainfall and higher humidity occurs regularly during the production seasons of most vegetables.

In addition, pesticide use is not adequately documented in most situations, and few states have coordinated pesticide use reporting procedures. Until 1990 when the State of California implemented total reporting of all pesticide applications, only uses of restricted-use (category 1) pesticides or applications made by commercial applicators were required to be reported. Yet this is one of the few databases that allows the magnitude of pesticide use on vegetable crops to be examined to some extent. Analysis of this database shows that in many cases few pesticides represent the greatest volume of usage on many crops, including those crops which appear to be fairly high recipients of pesticides. It is interesting to note that the timing of some of the highest volume applications including those of sulfur (in some cases), 1,3 dichloropropene and methyl bromide is not such that residues would occur on the harvested product. The relative amount of insecticides and fungicides used, in particular, will vary from region to region in the United States.

Pest monitoring programs, control action thresholds and predictive tools have been developed for certain pests and vegetable crop systems. When implemented, these programs have been demonstrated to reduce the need for conventional pesticide applications. Monitoring programs have the additional benefit of anticipating or better detecting when a pest problem could occur, thereby actually lowering the risk of damage for growers.

The development of monitoring guidelines and predictive tools for vegetable crops has been hindered relative to larger acreage crops because of the specificity of such programs to individual crops and regions. Further, little support for research to develop such programs has been available on a national basis.

Some examples of implemented programs include the use of sampling guidelines and control action thresholds for sweet corn, cabbage, onions and potatoes in New York and Massachusetts; broccoli and celery in California; and tomatoes in South Carolina. A sampling program for key worm pests of processing tomatoes in California has been demonstrated to have reduced insecticide use by over 40%, and has permitted the conservation of the parasitic wasp *Trichogramma* in at least one major production area. In Hawaii, a monitoring program and decision rules for the five major foliage pests of watermelon has permitted the conservation of leafminer parasitoids through reduced insecticide usage and the use of corn as a trap crop to lure melon flies.

Recently, processors including Campbell's Inc. and the San Tomo Group, among others, have actually implemented intensive company sponsored IPM programs for certain crops for which they contract in an effort to reduce pesticide inputs. They indicate an eagerness to implement other such programs if they were available. Del Monte has proposed sharing the cost of sponsoring Integrated Pest Management (IPM) research on specific crops with producers with whom they contract.

Computer-based predictive tools such as the Envirocaster is used for field onion and potato disease forecasting in Michigan; and PCM, the potato crop management program, is used by growers in Wisconsin to manage irrigation, early blight, late blight and selected insects. Various phenology or degree-day models are used for specific insect pests to time pesticide treatments, preventing unwarranted applications.

What is especially interesting about most of these systems is that better monitoring or predictive tools have permitted the integration of biological alternatives for controlling pests because the pest populations are not allowed to increase to the point where only conventional materials would be effective. In addition, these information based programs dramatically reduce the number of unnecessary pesticide applications.

Alternative pest management approaches are used to varying degrees as substitutes for conventional pesticides for specific pests and vegetable crops in IPM programs. Biological controls which include the conservation of beneficial organisms, inundative releases of beneficial organisms, the use of microbial agents and the introduction of exotic biological control agents are among the approaches used for specific crops. Resistant and pest-tolerant cultivars are available for several vegetable pests, and when available are widely utilized. Cultural practices that affect the life cycles of pest species, or reduce pest populations are also important methods of controlling many pest species.

Management of lettuce mosaic virus is an excellent example of a program relying on nonchemical alternatives to control a devastating pest species. In the 1950s and 1960s, lettuce mosaic virus, which can be vectored by aphids but also transmitted through seed, threatened lettuce production in the United States. An integrated approach was developed for its control which was based upon resistant cultivars, a testing program which produced certified "clean" seed, weed reservoir control, vector control, rapid plowdown of old lettuce fields and extended lettuce-free periods and zones. Today, these methods are universally adopted, and the disease is well under control.

ALTERNATIVE CONTROL TACTICS - STATUS AND CURRENT RESEARCH

As mentioned, biologically oriented IPM alternatives are available and being used for some specific pest and crop situations. In addition, other alternatives that are available, but not widely used, may become adopted as the availability of pesticides currently used for control are lost and the economics of pest control changes. New technologies in biological alternatives, cultural controls and biorational pesticides are being researched, and additional control tactics could become available. In fact, it seems entirely possible that potential alternatives exist or could be developed for many, if not most, pest problems afflicting vegetable crops.

However, it would be wise to be cautious in assessing the future of biologically intensive IPM for vegetables as it is impossible to predict the economic consequences of converting to a system of managing pests with severe restrictions on the availability of conventional pesticides. It is also impossible to develop an accurate timetable for development of such alternatives. Research on some pests,

113

such as certain seedling diseases in several vegetable crops, has been ongoing for decades without complete success. In some cases, even employing pesticides for control has not been entirely satisfactory.

Alternatives to pesticides are often quite specific and must be individually developed for each pest and crop situation. For instance, most biological control agents are effective against only one pest or a small group of pests, often in a geographically limited area. The narrow range of affected organisms is one reason why biological alternatives are more environmentally sound than broad-spectrum pesticides, but it also accounts for their limited development and use. Not all biological controls are specific, however. A recently registered strain of *Trichoderma hauzianum* is widely adapted to various crops, soils and pathogens. It can be applied in a seed treatment mixture that gives this bioprotectant an early advantage over competitive microflora.

Vegetables, in particular, are grown to some extent under a variety of different conditions in most areas of the United States. Growers may be faced with controlling a dozen or more major pests in any given vegetable crop, and these typically range from insects and plant pathogens to nematodes and weeds. Providing an alternative for a single pest won't necessarily lead to an overall reduction in conventional pesticides unless it is part of an integrated pest management program that considers appropriate and nondisruptive ways to manage other pests as well. There is a need to develop ecologically-based, crop-oriented integrated pest management programs that will become part of the overall production system. Various types of alternative approaches could be used alone or in combination in developing such IPM systems.

Classical Biological Control With Arthropod Parasitoids or Predators

Classical biological control involves the deliberate introduction and establishment of natural enemies into areas where they did not previously occur. Classical biological control programs are employed largely against pests of exotic origin that have become accidentally established in new areas without their key natural enemies. Recent examples of classical biological control for a vegetable crop include the introduction of the parasitoid *Trissolcus basalis* into California tomatoes to control the southern green stinkbug, an introduced species.

The parasitoid *Cotesia plutellae* is being introduced as a potential control for the diamondback moth in cole crops, and parasitoids are being introduced for the control of both the asparagus aphid and the

114

sweetpotato or cotton whitefly which have only recently become introduced into vegetables in the United States.

This classical approach has primarily targeted insect and weed pests, where it is possible to achieve high levels of control. Biological control agents of plant pathogens do not usually persist from season to season in sufficiently high levels to make this approach feasible.

Conservation

In many instances, biological agents occur in crop systems which can dramatically affect pest populations. These agents can be seriously impacted by the use of nonselective pesticides. Methods of conserving beneficial organisms include the use of control approaches which are not harmful to the beneficial organisms, or mechanisms for preserving biological control agents in other ways. Restricting the use of broad-spectrum pesticides prior to the development of a marketable product is a simple example. In Hawaii, watermelon growers have reduced pesticide sprays for leafminers by 90% through the intentional conservation of leafminer parasitoids including *Chrysonotomyia punctiventris* and *Ganaspidium utilis*.

Increasing our knowledge of the impact of native natural enemies on key pest species is essential in order to develop conservation or enhancement programs. For example, it is well known that native natural enemies such as *Diadegma insulare* and *Microplitis plutellae* exist for the diamondback moth, and that *Hyposoter exiguae* commonly parasitizes the beet armyworm. It is unfortunate that their true potential to control their respective key pest species is not well understood because of the treatments commonly applied to the vegetable hosts of the pests in a given region.

Augmentative Releases of Arthropod Parasitoids or Predators

Augmentative releases involve mass rearing of natural enemies or other biological control agents in a rearing facility, and release in pest-infested fields. Unlike classical biological control, it is assumed that regular releases of large numbers of biological control agents will be required to adequately manage the pest. Predaceous or parasitic insects, mites and nematodes have been used in this manner. Except for a few commercially available species, however, mass rearing methodology and facilities must be developed to make such systems economically viable. Some parasitoids such as *Trichogramma* and *Encarsia* are available from commercial insectaries, and have been shown to be useful in controlling Lepidoptera pests and whiteflies in field experiments and commercially in greenhouses.

Lacewings and predaceous mites are also available from commercial insectaries, and have found specialized crop niches. Predaceous mites, for example, are released on about half of California's strawberry acreage each year for control of the two-spotted spider mite. Although they are much more expensive than conventional pesticides, some growers believe that they are a useful addition to an integrated control approach. Entomophagous nematodes are also commercially available, and have found use in some vegetable crops particularly against soil-infesting insects. A few sweetpotato growers in southern Florida now use them as their primary tool for control of the sweetpotato weevil.

Biological control agents that have proven successful in some environments may be able to be made useful in other environments through genetic improvement. Genetic improvement could include modification of genes so the biological agents could survive pesticide treatments or a more harsh climate. Biological agents might also be selected to attack other pest species. Predaceous mites have been selected for resistance to several common insecticides, and can be successfully used with those materials in an integrated program. Recently, several geographic strains of the leafminer parasitoid *Diglyphus begini* were discovered which exhibit resistance to pyrethroid insecticides, and efforts are underway to mass rear the strain and disperse them to celery and tomatoes grown at other locations. A second leafminer parasite, *Ganaspidium utilis*, is being laboratory selected for pyrethroid resistance.

Cover Crops, Living Mulches and Trap Crops

Cover crops are noncrop species (resident or planted) grown to reduce undesirable pest populations or to provide other benefits. Cover crops may also be used as a mulch (either living or killed) into which annual crops may be planted with reduced tillage or nontillage. Weeds may be suppressed through physical competition or allelopathy. Living mulches block light and prevent germination of photoblastic weeds. Nematodes may be suppressed through antagonistic or allelopathic effects or by preferential attraction to the cover crop (also called trap crop) over the desired crop. Cover crops also may provide habitat for natural enemies of insects, mites and other pests. Similarly, the effectiveness of native natural enemies and other biological control agents can be improved by providing adjacent areas to improve shelter or food.

In New York, sweet corn grown through a clover sod has been shown to inhibit weed growth as well as provide control of the fall armyworm. Straw mulches used in potato production have been shown to provide a measure of control for the Colorado potato beetle, particularly when integrated with other approaches. However, logistics of implementing such practices on a large scale would have to be devised.

Cover crops in particular are promoted today as part of low-input sustainable systems for the benefits they provide particularly in increasing soil fertility and structure. Their impact upon pests and associated beneficial organisms, however, is not well understood, and much research is needed to use them predictably for pest control.

Microbial Agents

Microbial agents include bacteria, fungi, viruses, protozoa or nematodes that are pathogenic or antagonistic to pest species including insects, plant pathogens, nematodes or weeds. Some of these may occur naturally in the field, but most research is directed at those that can be mass produced and applied as a pesticide. In some cases, it is in fact the toxins synthesized by microbial agents that are actually applied to control pests. Because microbial agents can be chosen so they selectively kill only certain groups of pests, they are much safer for humans and wildlife than conventional pesticides.

Bacillus thuringiensis is the most widely known microbial agent for control of insects, particularly larvae of Lepidoptera. Although commercially available for many years, it has not been regarded as particularly effective when compared to conventional insecticides. In part this has been due to its specificity, unique mode of action and short residual activity. These properties are attractive to some vegetable growers today, particularly those who are concerned about residue on harvested products. Others use it as part of an integrated pest management program to decrease the selection pressure from use of conventional pesticides for control of such pests as diamondback moth and cabbage looper on cole crops, lettuce and tomato. The quality and level of *B. thuringiensis* production has improved dramatically, and today's products provide a predictable level of control and are competitively priced with many conventional pesticides. However, monitoring of pest populations becomes especially important when applying this material as it is most effective against younger larvae and at lower population levels. New isolates of *B. thuringiensis* have been identified, and one in particular is commercially available to control Colorado potato beetle on potato, eggplant and tomato. Recently, a commercial product of genetically altered *Pseudomonas fluorescens* containing *B. thuringiensis* delta endotoxin-producing genes has become available which is intended to increase the residual activity of the material by screening ultraviolet light which breaks down the toxin.

Viruses are known which attack key pest species of vegetable crops. For example, a nuclear polyhedrosis virus specific to *Heliothis zea*, the tomato fruitworm or corn earworm, has been registered for use on several crops and awaits production by its registrant. Research is currently underway on the use of a beet armyworm virus together with *B. thuringiensis* to control Lepidoptera pests of tomatoes in California.

Microbial agents hold promise for control of plant pathogens and possibly weeds as well. A commercial product containing *Bacillus subtilis* is available on snap beans for control of root and hypocotyl rots. Biological seed treatment using *Trichoderma* is available for some uses in controlling soil-borne diseases such as *Fusarium*. Research on other such treatments is also underway. A *Gliocladium* strain has been registered for use as an amendment for greenhouse planting mixes.

Successful use of microbial agents depends upon a highly effective strain, a production system that gives rise to high levels of the bioprotectant or biomass in sufficient quantity to be dried with high levels of active propagules and good shelf life. Finally, delivery and formulation processes and systems must permit the bioprotectant to grow rapidly in its new environment, even in the presence of large numbers of competitive microorganisms.

Research on formulations of microbial agents is essential for their widespread applicability. Addition of C-N food base, ultraviolet and infrared protection and pH control can increase residual activity and control.

Host Plant Resistance

Host plant resistance involves the development of crop cultivars that are able to resist or tolerate attack by pests that kill or seriously damage other plants of that species. In the case of weed pests, resistant cultivars would have improved ability to compete with weeds either through physical or allelopathic means. Once developed, host plant resistance provides a very inexpensive, environmentally sound, and highly effective way to avoid pest problems.

Host plant resistance for plant pathogens, nematodes and insects has been an important focus of research by plant geneticists and pest managers for many years. However, most emphasis has been on breeding cultivars resistant, or tolerant of pests for which no chemical alternatives were available or for which controls were costly. Selected examples of diseases commonly controlled by resistant or tolerant cultivars include downy mildew of lettuce and crucifers, *Fusarium* (races 2 and 3), *Verticillium*, tomato spotted wilt and tobacco mosaic virus of tomato, cucumber mosaic virus of cucumber, white rust of spinach, celery *Fusarium* yellows, white mold of beans, root rots of peas and tobacco mosaic virus of pepper. Some of these resistant crop cultivars are appropriate for certain growing regions only. Thrips resistant cabbage cultivars are available, as are corn cultivars which provide better tip coverage to prevent entry of insect pests. Nematode resistant tomato cultivars have become widely used in California.

Crop Rotation and Fallow Periods

Crop rotation involves the deliberate planting of specific crop sequences to make year-to-year survival of pests, particularly those which feed upon a narrow range of plant hosts, difficult or impossible. Crop rotation can provide good control for certain fairly host-specific plant pathogens, nematodes and insects. Rotation can also contribute to weed management by planting particularly competitive crop species, or ones which can be controlled using selective herbicides. Fallow periods also permit the interruption of pest population cycles by removing possible crop hosts.

Crop rotations and fallow periods have lost favor in some vegetable production areas such as California and Florida where increasing land values have mandated that high value vegetable production be continuously practiced. It is interesting that the recent development of nematode-resistant tomatoes has permitted their use in rotation with high value crops for which resistant cultivars are not available. It seems possible that breeding programs targeting crops which could be used in such a rotation would hold promise.

Physical Controls

Physical control methods include those that mechanically destroy pests or result in physical barriers to pest infestation or movement. A large variety of methods could be included in this category. Some are regular farm practices such as cultivation, while others such as solarization, artificial mulches, row covers, sticky barriers, flaming, and suction machines are applied specifically for pest control. Solarization is being used to some extent for control of soil-borne diseases and weeds in certain specialized situation such as sweet potatoes in California. Improvements in application, biodegradability and lower costs resulting from commercial efforts now being pursued may make this a useful approach for many high value crops such as vegetables. Plastic row covers which exclude virus vectors and increase temperatures to control some diseases are used in the production of cucurbits in the desert valleys of southern California. Hot water treatments are commonly used to control nematodes in garlic and strawberry planting stock. Suction machines are used by California strawberry growers to control early season *Lygus* bugs without pesticide treatments.

Sanitation

Sanitation involves any practice that removes sources of inoculum or removes overwintering or alternate refuges where pests can survive when the crop is not in the field including both crop and noncrop hosts. Cleaning equipment to avoid transmission of inoculum is an excellent example. Many sanitation methods may be considered just plain "good farming" practices which are already widely utilized such as weed removal and destruction of crop debris.

Florida tomato and pepper growers help control the spread of bacterial spot by regulating the activity of field workers. In transplant production, hand washing stations are provided in each greenhouse, and workers are assigned to "clean" jobs to begin the day before they do "dirty" jobs such as roguing or cleaning returned flats. During transplant production and for the first several weeks after plant-out, the removal and destruction of diseased plants is a key strategy in managing bacterial spot.

Water Management

Certain changes in irrigation management can reduce or increase damage by insects, mites or plant pathogens and can impact weed competitiveness. Changing methods of water application from flood to furrow to sprinkler to drip can have broad impacts. Subsurface drip irrigation is being tested in some areas where rainfall does not occur during the production season. This system combined with transplants of plants such as tomatoes can allow the crop plant's root system to receive moisture while the soil surface remains dry preventing weed germination. Providing good soil drainage when preparing fields and planting onto berms can provide some measure of control for certain pests in special situations.

Preplant Decisions

Decisions a grower makes at planting time can have a significant impact on future pest problems. These decisions can include choice of field (soil type, previous crops, existing pest populations, etc.), location of field relative to sources of pest infestation or inoculum, time of planting or harvest to avoid major pest migrations or unfavorable weather conditions which are conducive to pest development, planting density, transplanting versus direct seeding, row spacing to allow air movement, etc.

Regulation

Mandatory host-free periods, host-free zones or crop termination dates, seed indexing and certification of pest-free stock have been used effectively to control several pests on a regional or commodity-wide basis. Western celery mosaic virus has been controlled in California and Florida by a host-free period. Pest detection and exclusion are also important government regulated activities.

Horticultural Oils and Soaps

Horticultural oils, including highly refined, food-grade crop oils, and various fatty acids are known to have insecticidal and miticidal properties, although they are generally regarded as less effective than conventional pesticides. These materials have less impact on human health and wildlife than conventional pesticides when properly applied. Fungicidal and herbicidal soaps may also have some potential.

Botanicals

Botanical insecticides are those which are extracted or derived directly from plants. Common botanicals include pyrethrum, neem, rotenone, sabadilla, ryania, limonene and nicotine. Most of these materials have broad-spectrum activity, and some are acutely toxic to humans and wildlife. Because they break down rapidly in the environment, many consider them safer than synthetic organic materials. Some such as neem and limonene have antifeedant properties.

Semiochemicals

Semiochemicals are chemicals produced by insects which affect the behavior of other insects. Allomones and kairomones are substances given off by a plant or animal species that causes a specific behavioral response in another species. Allomones (such as repellents, deterrents, or antibiotics) favor the producer. Kairomones (such as attractants or feeding stimulants) favor the receiving organisms. However, either type could be used in a pest management program. Pheromones are chemicals given off by individuals of one species that elicit specific behavior by other individuals of the same species. The ones most commonly used in pest management are sex attractants that attract individuals of the opposite sex for mating. They have been demonstrated to control certain species through mating disruption and less commonly by mass trapping. Insects for which pheromones are registered as pest controls include the artichoke plume moth and the tomato pinworm. Both pests are being controlled commercially to some extent using pheromones.

Recent studies have demonstrated that improved dispenser technology can increase the efficiency of semiochemicals in controlling pest species. Yet other studies have shown that volatiles produced by plants can improve the attractiveness of semiochemicals, or can by themselves provide responses which could be used both in population monitoring and pest control. Noctuid moths, which include some of the most significant insect pests of vegetable crops, have been targets of many of these studies.

POSTHARVEST HANDLING

Most control programs for postharvest pest problems involve several individual practices that begin with harvest and conclude with marketing. There is rarely a single practice that completely prevents postharvest diseases in particular, although refrigeration appears adequate for certain ones. Therefore, most strategies for postharvest control require an integrated pest management approach.

The diversity of pest problems responsible for postharvest losses are few in comparison with those found in field or greenhouse crops, but in many cases they are more difficult to control. For example, most postharvest pathogens can exist as aggressive and successful saprophytes. They may be dispersed long distances (e.g., recent reports about the soft-rot erwinias), and they quickly colonize stressed or senescent plants such as those which are mature or stressed by having been harvested. Finally, fresh vegetables must be kept in a succulent condition, and they are stored under high humidity which helps promote pathogen development. This also precludes practices that are effective for control of postharvest deterioration of grains and other dry foods and feeds. Postharvest handling of vegetables intended for processing has other unique problems.

For most fresh vegetables, cold storage, moisture control and sanitation are primary methods of control. Biological agents are not being used to control postharvest diseases of vegetables at present, but they hold some promise for the future. Delivery of effective numbers of antagonists to potential infection sites has been a limiting factor. Typically, antagonists must be present in 10 to 100-fold greater numbers than the target microbe for effective control. This would be exceedingly difficult in commercial practice. Current research on biological control of soft-rot erwinias focus upon virulent bacteriophage and pyrrolnitrin which is produced by *Pseudomonas cepacia*. In addition, the genetic controls for enzyme production and export have been found and defined.

The ecology of the soft-rot erwinias is not well understood. Many different strains of soft-rot erwinias have been described, but the potential for the various strains to cause losses is unknown. For example in potatoes, five or six of more than 40 serotypes have been consistently associated with major outbreaks of bacterial soft-rots while the role of the other serotypes and of the strains that cannot be serotyped is unknown. The ecology of other postharvest pathogens is less understood than that of the soft-rot erwinias. With this lack of knowledge of the ecology of postharvest pathogens, it is understandable that biological control researchers are reduced to screening numerous naturally and artificially derived microbes for activity against pathogens. The lack of knowledge of pathogen ecology is a technical constraint to development of an effective IPM strategy. Support for IPM, research on pathogen ecology can be stimulated by funding. If biological control organisms are identified which appear to be effective in postharvest systems, several constraints must be overcome. Economic fermentation culture of immense quantities of microbes would be required for control. In addition, some yeasts being studied may be related to human pathogens. How people with suppressed immune systems respond to high populations of microbes on their produce should be a source of concern.

FUTURE RESEARCH ON BIOLOGICALLY BASED ALTERNATIVES

Research is currently being conducted for selected pests using each of the biologically oriented integrated pest management approaches mentioned previously, and the possible applications for such research on other vegetable crops and associated pests is virtually unlimited. In order to appreciate the targets for future research, a comprehensive review of each vegetable crop system would be necessary by region. There are many significant pest problems among vegetable crops, and many are not being addressed due at least in part to lack of research staff and inadequate funding.

Critical areas of future research mentioned by researchers responding to our survey included such general areas as:

- Increased knowledge of the basic biology and field ecology of both pest and beneficial organisms including microbial agents and antagonists, their interactions and their effect on crop production.

- Increased emphasis on epidemiological studies of pests, and forecasting systems.

- Development of practical sampling techniques for pests.

- Improved methodology for increasing beneficial organisms in culture and augmenting natural populations.

- Identification of selective pesticides which are "soft" on beneficial organisms.

- Surveys for pesticide-resistant natural enemy strains and/or laboratory selection for resistant natural enemies.

- Increased emphasis on nonchemical methods of weed control which are economically viable including living and dead mulches, electronically-directed cultivation and bioherbicides.

- Increased support of research on the ecology of plant associated microorganisms.

- New or modified control action thresholds which incorporate population levels of beneficial organisms.

- Increased emphasis on identification of genetic resistance or tolerance for all major vegetable crop systems, and development of mechanisms to prevent pest resistance to resistant cultivars.

- Increased efforts in exploration for new species or strains of beneficial organisms.

- Isolation of new beneficial microbial agents or antagonists, and enhancement of those already identified to improve activity.

- Renewed support of directed, on-farm adaptive research.

- Support for biologically intensive IPM research specifically addressing pest problems of the food industry.

- Research on the effects of various crop rotations on pest populations.

- Development of formulations and delivery systems permitting reliable performance by effective microbial agents.

COOPERATIVE EXTENSION'S ROLE

The role of Cooperative Extension in all areas of agricultural research and education has changed over the past decade or more. This has become especially true in pest management, where a small but increasingly significant and well-educated cadre of private pest control advisors have begun to provide advice to growers on the efficient use of pesticides and biologically oriented pest management practices. Cooperative Extension in many states has taken a leadership role in identifying the need for such individuals, demonstrating their impact to growers and providing continuing education training to the pest control advisors in new integrated pest management strategies and tactics. In other states where a viable private sector has yet to develop, scouting programs provide a similar service to growers.

The traditional role of Cooperative Extension is in the validation of research and the transmission of information to and from researchers and growers. In many cases, this work has provided an independent, third-party evaluation of new technology, and has allowed the adaptation of new technology to specific regions and crops. This continues to be important to integrated pest management programs which are increasingly more complex and information-intensive. In many states, Cooperative Extension has assumed the important role of conducting "adaptive" or "applied" research in integrated pest management, and has begun to bridge the gap between fundamental research and its application. Field demonstration of IPM techniques continues to be a critical educational service provided by Cooperative Extension. In many cases, growers and their consultants need to see with their own eyes and experience with their own hands the concept of IPM in order to embrace it.

The critical role of Cooperative Extension in the adaption and implementation of integrated pest management has been well documented in several studies, and it was generally agreed by respondents to our survey that increased demands placed upon Cooperative Extension staff coupled with reduced budgets has made the process of both adaptive research and technology transfer more difficult. Some states, notably New York, Texas and California, have made significant investments in formally identified Cooperative Extension integrated pest management programs.

CONSTRAINTS TO THE USE OF BIOLOGICALLY ORIENTED IPM SYSTEMS IN VEGETABLE CROPS

Integrated pest management has received much attention, particularly in the last decade, as the only rational approach to providing long-term solutions to pest problems in vegetable crop production. When successfully implemented, integrated pest management programs and biologically oriented pest control tactics have been demonstrated to provide benefits such as reduced risk of pest damage, increased net profits to growers and reduced pesticide use. Yet IPM systems and practices, even when available and demonstrated to be effective, are far from being universally adopted. A number of obstacles to adoption have been identified including those which can be loosely categorized as technical, economic and institutional. The impact of governmental policies and regulations will be discussed later.

Technical Constraints

The development of IPM strategies and biologically oriented pest management techniques that may be applied to vegetable crop systems is labor-intensive and can take years to develop for specific applications. Some crops such as potatoes and tomatoes are fairly well-studied in some growing regions, and multiple pest management tactics are being employed for several target pest species which has resulted in a reduction in the amount of conventional pesticides required for control. In these crops and regions it may be possible to move toward even more biologically intensive systems involving pest and beneficial monitoring, host plant resistance, utilization of natural pest controls and other techniques rather quickly. In other growing regions and in the many other vegetable crops which are less well studied, there is often a basic lack of understanding of both crop and pest biology as well as their interactions which precludes the use of integrated pest management in general. This lack of understanding typically reflects a lack of research effort, or adaptive studies. Research on biologically oriented pest management strategies and their specific applications requires adequate levels of support, and only in rare instances is there support for research on most vegetable crops from either government or industry sources. The need to simplify IPM methodology is also especially important, particularly with respect to developing monitoring and sampling guidelines for pests and beneficial organisms. Crop consultants labor under time constraints, so it is essential that economic considerations for implementation receive the same consideration as statistical accuracy in their development.

Economic Considerations

Vegetable crops typically have a high unit value which can affect the production and marketing objectives of a grower. The value of a relatively small loss in production can be quite high, and the cost of using an input such as a conventional pesticide is low relative to crop value. In addition, there is a perception widely held by growers and others that IPM is risky and does not offer short-term economic advantages compared to conventional control, particularly because of additional labor costs from sampling and monitoring, and less effective control tactics. In short, pesticides are perceived to be cheap and easy to use in spite of increasing costs and recent regulations on their use and application. They are relatively easy to obtain and their sources are reliable. In addition, pesticides give growers immediate reinforcement in terms of pest control. Therefore, most growers have developed confidence in their use. Growers must be convinced that IPM advice has value, and that in some cases substituting biologically intensive methods is actually beneficial. In addition, the cost of vegetables to the consumer is relatively low, and the profit margin for the grower relative to this cost is also low. Therefore, there is little potential for the grower to voluntarily substitute a more expensive but less environmentally disturbing control technique for the use of conventional pesticides. A relatively modest increase in food prices passed along to the grower could facilitate the change to alternative control approaches. Better economic analysis of biologically oriented approaches is important in this regard. The development of more biologically oriented tactics specifically applicable to vegetable crops is important to increase the visibility of such alternatives.

An IPM strategy such as monitoring is a tool growers could use for managing risk; as the more growers learn about pests in their fields, the lower the likelihood of resulting damage. This has been documented in California processing tomatoes where a monitoring program for Lepidoptera larvae has been shown to reduce both the risk of damage and the amount of pesticide required for control.

Lack of funds for both research and extension programs for vegetables is a major factor in the relative inattention to integrated pest management and biologically oriented controls on those crops. Most federal funding for research and extension activities has been targeted toward farm program crops such as cotton, soybeans, corn, etc., rather than consumer noticeable crops which are of comparatively lower acreage and are geographically dispersed.

Institutional Constraints

The lack of support for interdisciplinary collaboration in pest management research, extension and teaching has been a major constraint raised by respondents to our survey. Some questioned what they perceive to be the low priority given IPM within our universities. There is a tendency for research and educational activities at the Land Grant Universities to be conducted within strong discipline-oriented departmental units (e.g., agronomy, plant pathology, entomology, etc.) which have evolved in response to institutional pressures for specialization. Individual achievements rather than collaborative accomplishments are typically encouraged by evaluation criteria, leading to the predominance of publications and research grant proposals which stress a narrow focus at the expense of true cropping system studies involving biological, social and economic components. There is little incentive for multiple ownership of intellectual achievements. Even federal research grants programs require an individual to be designated the principle investigator. Similarly, degree programs are structured to produce graduates with narrow disciplinary expertise rather than general knowledge of cropping systems.

Institutional communications channels for biologically oriented management tactics are not well developed. Growers receive pest management information from a variety of sources. In this regard, chemical controls have a competitive advantage over their biological or nonchemical counterparts. There is a well established infrastructure for information and promotion of conventional pesticides, with a high ratio of chemical sales personnel, technical representatives, and farm supply dealers to private pest management consultants and extension IPM staff. In addition, many alternative approaches are nonproprietary, and will likely never be produced or marketed through corporations providing traditional agricultural services. Even biologically oriented products which may be proprietary are often available only locally or through venture capital corporations which have not established adequate marketing channels in the agricultural community. An exception is the production and marketing of *B. thuringiensis* products by several major corporations.

IMPACT OF FEDERAL AND STATE POLICIES AND REGULATIONS

It is essential that integrated pest management and increased emphasis on biologically oriented controls become institutional policy in governmental agencies. Coordination is necessary to avoid conflicting policies or programs within an agency. It was stated, for example, that there is too much emphasis within the USDA on program crops, those which receive federal price supports such as corn and cotton. Emphasis on research and extension efforts related to program crops has actually hindered the development of biologically oriented alternatives for vegetable crops by diverting research grant support and therefore researchers from work on vegetables. Crop insurance which protects producers of program crops should be extended to those who grow vegetables as well. Insuring vegetable growers who use biologically intensive approaches to pest management from pest losses would increase use of alternatives. Marketing orders which use low pest damage standards as a tool for regulating supply are another perceived inducement for continued pesticide use. Farm subsidy programs including commodity price supports encourage growers to plant the same crop each year to qualify for benefits. This discourages crop rotation and diversification which are effective methods for controlling many species of insects, diseases, nematodes and weeds. A few survey respondents felt that IPM research should be expanded even beyond crops to all segments of the food industry, with industry groups such as the National Food Processors Association assisting in identification of needs.

Quality standards imposed by government agencies such as the Federal Food and Drug Administration, the United States Department of Agriculture and state departments of agriculture impact the ability for applying many biologically oriented alternatives in vegetable crops both in the field where extremely low damage levels may be necessary to meet quality tolerances and in postharvest handling and storage where diseases and insects directly impact quality and product life. An extreme example is that federal policy does not allow the use of insect parasitoids in areas where food is stored as packaged, finished products, where the parasitoids could not become contaminants. Although products with cosmetic blemishes can be just as fresh, nutritious and tasty as products without blemishes, unblemished products are likely to remain fresher in storage. Moreover, blemishes can become colonized by mycotoxin producing fungi or be contaminated with stress metabolites produced by the plant. Where issues of human health are concerned, there is no question that such regulation is necessary. However, when quality standards have been imposed in response to consumer concerns, there needs to be a mechanism for establishing tolerances based on the input of consumers and processors or packers who also consider issues such as pesticide residues on consumer noticeable crops such as vegetables.

Plantback restrictions that are placed on pesticide labels should not inhibit good pest management practices such as crop rotations. This is particularly a problem on vegetable crops for which few pesticides are labelled for use. For example, dry beans cannot follow potatoes that have been sprayed with chlorothalonil, yet snap beans, which are the same species, can be grown.

Many respondents to our survey indicated that the application of recent advances in biologically intensive research is being stifled by inadequate regulation and registration procedures. Examples mentioned included the lack of clear, consistent regulation of microbial pesticides, particularly those which are genetically modified. They cited the difficulties experienced with field tests of "ice minus" and recombinant *Pseudomonas* in the mid-1980s as examples. Lack of established criteria for field tests involving genetically engineered microbes was also mentioned as a problem. Such regulatory hurdles are seen as preventing commercial development as companies are unwilling to invest in products which have an uncertain regulatory future. Consideration should be given to generic registration of microbials and semiochemicals across broad categories of vegetable crops. Rules on interstate transport of indigenous non-pathogens and introductions of novel biocontrol agents was also seen as a hindrance to biological control programs.

POTENTIAL REGULATORY AND POLICY OPTIONS

Although many research and extension professionals are involved in integrated pest management activities including the development of biologically intensive pest management tactics in Land Grant Universities, the USDA and private companies, the level of activity is insufficient to meet the demands that would be necessary to develop and implement programs universally in the United States. This is especially true of vegetable crops where each cropping system and production region should receive intensive study. The obvious response as mentioned by many respondents would be to direct more funding to those researchers involved in such activities. In fact, funding levels for applied ecological studies such as those required to develop and implement biologically intensive systems on vegetable crops has declined over the past decade. As state support for this work has declined, external funding from industry and more importantly from federal agencies who support very basic research has increasingly leveraged what support remains, resulting in even less emphasis on applied ecological studies and biologically applied systems.

Greater investment in research and extension, while necessary, should be accompanied by mechanisms which would address both near term and long-term problems associated with conversion to an agricultural production system based on lower pesticide use. Research and extension funding programs should be more goal oriented, no matter what the time frame of the proposed research. Individuals designing and administering these programs should clearly identify the goals of the programs. Research, extension and the private sector should be closely linked in both the development and implementation of the programs.

Private sector incentives for developing and marketing biologically oriented controls should be made to interest more companies in development and production, and to hasten their use. This might take the form of grants or tax incentives. Mechanisms should be established for the rapid review of biological pesticides, including fungi, viruses and bacteria, and other alternative pesticides such as pheromones which are submitted for registration. Requirements for registration should be clear and consistent. Present registration rules which apply to conventional pesticides are not always appropriate for these new products. It is also unclear whether such products are seriously considered as alternatives to conventional products under special review. Closer linkage of research and regulation could improve the process. Regulators should be cautious, however, not to jeopardize public confidence in the process so that biologically intensive tactics do not meet a similar fate to conventional pesticides.

Some growers have attempted to use augmentative releases of commercially available biological control agents without success, and these growers are therefore unlikely to do so in the future. Many times the lack of success can be attributed to questions of quality or inappropriate use of the agents. Certifying commercial insectaries with respect to ethics and practices could be useful in assuring growers of quality products and adequate technical support in their use.

Because conventional pesticides are generally considered cheap relative to the value of many vegetable crops, the price of chemical pesticides could be increased to reflect their expense in terms of regulatory expense and social and environmental costs. User fees or taxes on conventional pesticides could be an option.

Implementing IPM is best achieved by individuals well trained in pest management and the plant sciences. The success of private pest management consultants in reducing pesticide use through implementing IPM strategies and techniques is well documented. However, the number of private pest management consultants is limited, and is virtually nonexistent in some regions of the United States. Increasing the utilization of private pest management consultants could be accomplished by increasing grower incentives, or by licensing these individuals and requiring their direct input in the pesticide use process. Degree programs which would lead to professional accreditation in plant health and pest management should be supported at Land Grant Universities in much the same way as are veterinary schools, and they could be structured as graduate groups or professional schools at those institutions. No comprehensive programs currently exist at any Land Grant institution, but some attempts have been made at both the undergraduate and masters degree levels. Plant health practitioners who would graduate from such programs, like their veterinary or medical counterparts, would be permitted to prescribe the use of restricted agricultural chemicals in response to careful monitoring of crop health and in accordance with label requirements.

Careful distinctions should be made between acceptable damage levels based upon health concerns or those necessary in order to maintain competitive products, and those that are strictly cosmetic in nature. Consumers and processors or packers should be consulted in the establishment of quality standards, which should consider pesticide residues and other issues. Consumer education is also important as quality standards largely reflect public desire for unblemished produce and processed products. Grading standards should emphasize nutritional quality rather than appearance, and a certain amount of blemish should be considered acceptable even at higher grading levels. In addition, harvest timing is often market-driven, and this will have an impact on grading standards.

Some survey respondents felt that government certification might be an incentive for consumers to purchase products grown using specified practices, and would recognize those growers utilizing such practices. Certification would require a set of crop-specific standards which a grower would need to follow in order to meet requirements. A few states define and certify organically grown products. However, it will probably be more difficult to define and regulate biologically intensive integrated pest management systems. Other survey respondents felt that government certification programs would not foster IPM implementation, because consumers mistrust government programs as they relate to food safety issues. They also noted that acceptable IPM practices would need to be identified for specific growing areas, and therefore might be unrealistic to develop. It might be possible for growers groups to insure compliance through examination boards or through requiring the use of private pest management consultants in making pesticide decisions.

RECOMMENDATIONS FOR ACHIEVING SHORT, INTERMEDIATE AND LONG-TERM GOALS

Develop alternative (non-pesticide, or low impact pesticide) tactics.

- Increase the number and efficacy of biological controls for the myriad of vegetable and vegetable production systems.

- Develop incentives for the production and distribution of biologically oriented controls.

- Improve mass rearing procedures and quality controls of biocontrols.

- Develop clear, consistent, appropriate registration procedures of biorational controls.

- Develop facilities and support for isolation and engineering of improved biocontrol agents.

- Investigate, develop, and implement cultural controls.

- Develop and employ pest-resistant, and pest-competitive vegetable cultivars.

- Develop and employ biorational pesticides.

- Develop and implement highly targeted "reduced-pesticide" tactics.

- Expand agricultural research to include all facets of the food industry including processing and packing.

Speed the development and implementation of comprehensive IPM programs.

- Employ appropriately designed decision support systems.

- Include economic and environmental consideration in analyses of strategies.

- Employ appropriate computer technology for communication and analysis.

- Facilitate the transfer of research results to implementation programs.

- Enhance implementation technologies and resources.

Enhance the efficiency of currently available conventional and biorational pesticides.

- Increase the number of appropriate thresholds for use in comprehensive programs.

- Develop time-efficient sampling and monitoring methodology.

- Increase the number and accuracy of forecasts.

- Identify and utilize strategies that delay or avoid pesticide resistance.

- Decrease reliance on pesticides as the only or first pest-suppression tactic.

- Improve application technology of both conventional pesticides and biological agents for both efficacy and lower environmental contamination.

Enhance incentives for grower adoption of IPM programs and strategies.

- Consider such practices as enabling a price advantage for vegetables produced under certified IPM programs.

- Make crop insurance available for growers enrolled in IPM programs.

- Alter grading standards for vegetables to reflect competitiveness rather than appearance and the perception of health risk.

- Tax certain pesticide uses (foreign as well as domestic).

- Make certain pesticides available by prescription only.

- Alter governmental programs that encourage maximum production, rather than optimal production.

Allow development of location-specific, highly specialized IPM programs for vegetable production.

- Encourage development of location-specific biological controls.

- Encourage highly location-specific uses of pesticides.

- Make available weather information which is as location-specific as possible.

Develop a basic understanding of the ecosystems supporting vegetable production.

- Support basic studies on pest ecology and epidemiology.

- Support studies on important interacting components of vegetable agroecosystems.

- Support studies of the implications of various regulations on long-term status of vegetable agroecosystems.

Facilitate the development of a professional group of plant protection specialists qualified to prescribe uses of certain pesticides.

- Support appropriate legislation for establishment of such a group.

- Aid in developing the goals and regulations for such a group.

- Support educational programs training such professionals.

- Develop incentives for the production and distribution of biologically oriented controls.

- Improve mass-rearing procedures and quality control of biological controls.

- Provide facilities and support for isolation and engineering of improved biological control organisms.

STRATEGIES FOR ACHIEVING GOALS

Establish a Grants Program

- To develop alternative tactics
- To enhance the efficiency of currently available conventional and biorational pesticides
- To develop basic understanding of vegetable ecosystems

A grants program should be developed to augment current USDA Cooperative States Research Service (CSRS) grant support in the area of IPM research and development. The new program should foster the investigation and development of promising and potential alternative controls, improved uses of conventional and biorational pesticides, and new understanding of vegetable agroecosystems. Biocontrol investigations should include techniques of mass-rearing procedures as well as the identification and evaluation of potential agents. Additionally, there should be support for development of facilities. Investigations of targeted "reduced-pesticide" tactics should be done. Identification and development of pest-competitive or -resistant cultivars should be supported. Investigations of biorational pesticides should be fostered. Innovative uses of established conventional and biorational pesticides should be supported in terms of development of thresholds, sampling methodologies, and disease forecasts. Strategies which employ multiple tactics and which delay selection for resistance should be investigated. Additionally, basic studies of vegetable agroecosystem ecology need to be conducted in the important vegetable production areas. These will include studies on basic pest biology, ecology and epidemiology; studies of important interactions within the vegetable agroecosystem; and studies of the implications of various pest management activities.

The current level of IPM competitive grants funding should be increased dramatically to foster the proposed studies. In addition to increases in the IPM regional grants program (USDA CSRS), there should be additional programs to support investigations in biological control, resistance management, host resistance, cultural control, vegetable agroecosystem interactions, application technology (for both conventional and alternative pesticides), and evaluation and validation studies. Within 5 years, amounts available for these programs should approach $5-20 million per year for each program (totalling about $90 million per year). Management of the grants program should be conducted by a process similar to that used for the USDA competitive grants program.

Develop Regional Centers for Pest Management

Centers of excellence which enable focus on the location-specific problems in vegetable pest management should be constructed. These centers will facilitate the coordinated administration and facilitation of pest management in a region. The major activities will be to speed the development and implementation of comprehensive IPM programs that are unique to specific locations. Such activities are dependent on thorough understanding of the interactions of organisms, physical factors and human activities in agroecosystems. Funding for regional centers should be through federal agencies and regions or states. Activities will include the development of appropriately designed decision support systems, implementation of effective communication technology and support for implementation programs.

The centers should specialize in approaches that are unique for that particular region, and could entail studies of application technology (for conventional pesticides and for biorational, microbial, or bio-pesticides). Location-specific uses of conventional or alternative pesticides, or biocontrols should be developed with the aid of such centers. Such centers should facilitate the investigation of physical, biotic and human activities in important agroecosystems. Physical (atmospheric, and edaphic) factors, biotic and human activities must be investigated to learn of the long-term implications of various pest management and production practices. Investigation of the benefits of fine-resolution weather forecasts needs to be done.

Alter Regulatory Procedures to Foster the Adoption of IPM

A series of steps should be taken to illustrate to growers the benefits of IPM approaches. Some pesticides should be available for use only as part of a bone fide IPM program. The EPA should make some pesticides available by prescription only. (see below for suggestions of the mechanisms for identifying persons who can prescribe such pesticides). Pesticides to be available by "prescription only" might be those that are likely to be banned. Prescriptions would be possible only in those programs that utilize an IPM approach. "Generic" justifications for certain IPM practices (biocontrol and cultural approaches) should become possible for registering such practices. The USDA should develop a process whereby IPM growers can be identified, and their produce can be marketed as having been grown with an IPM approach. Details of the identification will have to recognize the site-specific nature of pest management, and regional differences will be expected. Thus the criteria to identify an IPM grower in one part of the country are likely to be quite different from the criteria that identify such a grower in a different part of the country.

Foster the Development of an Association of IPM Professionals

This group of professionals will become the individuals who will make pesticide prescriptions, and will transfer IPM technology from research and development personnel to growers. Both the USDA and EPA should facilitate efforts by existing practitioners and professional societies to develop registries and certification programs. The USDA should take the lead in organizing a working group of pest management professionals representing all pest management professional societies to develop criteria for certification and registration. Another working group should aid Land Grant Universities in developing educational programs that would prepare students for careers as certified IPM professionals.

VEGETABLE IPM
EXAMPLE: POTATOES

Potatoes illustrate the importance of pest management typical in the production of the more than 60 commercial vegetables produced in the United States. Potatoes are produced in diverse agroecosystems with diverse problems for diverse markets. As is the case with all vegetables, pesticide availability for potatoes is generally not a high priority for the chemical industry or for the EPA because it represents a minor use. However, as is true for most vegetables, pesticide availability is crucial to the continued production of potatoes.

Potatoes are a crop which requires intensive pest management. Many tactics are currently used routinely, the most important of which is probably the use of certified seed to assure the suppression of many pests which are associated with seed tubers. Cultural and sanitation practices, as well as pesticides are used intensively. Potato pest management is constrained by the extremely rigid demands of markets (processors and buyers), competitiveness between production areas and by federal grading standards. Some effective pest management tactics are not currently applicable because of market, grading or processing requirements.

Potatoes are beset by many pest problems. More than 80 pests (diseases, insects, weeds and nematodes) can damage potatoes in the United States, and pesticides contribute to suppression programs for at least half of these. For these and the other pests, there have been significant increases in the efficiency of management practices during the past two decades. IPM programs in various parts of the country have enhanced the effectiveness with which pesticides and other tactics are used, and in some of these agroecosystems only 60-80% as much pesticide is now used compared to a decade ago.

IPM Possibilities Given Limited Pesticide Availability

In order to stimulate thinking for the planning of future IPM research, the extreme position of total elimination of all pesticides was considered. This situation is highly unlikely, but was used to stretch imaginations and stimulate creativity. It is quite likely that pesticides in minor crops will become increasingly restricted, and the assumption of total elimination was the extreme of that trend.

The implications to potato production of the elimination of all pesticides were assessed for 30 individual or groups of pests (or production problems). In the majority of cases, elimination of pesticides would cause large production problems in some areas.

139

For 11 of the most important pests, alternative tactics were identified, and the feasibilities of such tactics were estimated in terms of technical and operational consideration (Table 2). Additionally, if data gaps existed and more research was needed, the time required to develop implementable results was estimated (Table 2). Of the many tactics identified some hold high promise for contributing to successful IPM (i.e., host resistance), others were deemed less likely to contribute (i.e., vacuuming plants to remove Colorado potato beetles). For the other 19 pests or pest groups only the relative impact of completely eliminating pesticides as a tactic was estimated (Table 3).

The results clearly indicate that pest problems are highly diverse in different regions of the country; pesticides are currently an integral and necessary component of potato production; and IPM is highly site-specific. Abrupt removal of pesticides would devastate the United States potato industry. However, there are many possibilities for reducing the magnitude of the role of pesticides, although use of these possibilities requires significant research and development. Some efficiencies of pesticide use are currently possible but need additional effort for implementation. Factors that aid implementation are considered next.

Table 2. Projected impact of elimination of all pesticides for selected potato pests in various regions of the United States, and identification of potential alternatives with estimates of research effort needed to implement the alternatives.

| Regional Importance w/o Pesticides[a] | | | | Pest | Alternative | Impl.[b] | | Research Timeframe[c] | | | Comment |
NE	C	W	S			T	O	1-5	5-10	10+	
4	4	4	2	Aphids	Biocontrol	X			X		
					Alarm pheromone	X		X	X		
					Resistance	X		X	X	X	
					Destroy alt. host	X	X	X			
					Stylet oils	X	X	X			Regionally applicable (for non-persistent viruses)
					Reflective mulch	X			X		Economical? Uncertain efficacy. Harvest problems
					Soaps	X	X				Expense? phytotoxicity
					Horticultural oils	X	X				Economics, phytotoxicity
4	4	4	2	Virus	Stylet oils (PVY)	X	X	X			Esp for seed production
					Resistance				X	X	Genetic engr, & tradl breeding
					Certified seed	X	X				Especially important

[a] Regions of the country are Northeast (NE); Central (C); West (W); and South (S); 4 = large impact of eliminating pesticides; 0 = no impact of eliminating pesticides

[b] Status of implementation T = technically possible; O = operationally possible

[c] Years to point of field implementation

Table 2 (continued).

Regional Importance w/o Pesticides[a] NE	C	W	S	Pest	Alternative	Impl.[b] T	O	Research Timeframe[c] 1-5	5-10	10+	Comment
2	4	4	4	Nematodes	Rotation	X	X				Expensive, species specific
					Host resistance	X	X	X	X	X	Currently very limited, species specific
					Microbials				X	X	Not economically feasible now
					Cover crops	X		X	X		
					Biocontrol		X	X	X		
					Solarization	X		X	X		Expensive, region specific
					Organic amendments						
4	3	3	3	Weeds	Plant density	X		X			Interrelationships to prod'n practices
					Cover crops	X			X	X	Economics, impact on other production practices
					Living mulches						
					Straw mulch				X	X	Economics?
					More cultivation	X	X		X	X	Economics?
					Hand hoeing	X	X				Economics
					Biocontrols					X	
					Solarization	X	X	X	X		Region specific
					Competitive cultivars	X	X	X	X	X	Limited selection currently
					Monitoring/prediction			X	X	X	
					Irrigation (subsurface drip)				X	X	

Table 2 (continued).

NE	C	W	S	Pest	Alternative	T	O	1-5	5-10	10+	Comment
4	4	1	1	Late Blight (P. infestans)	Sanitation	X	X				Currently used
					Resistance	X	X	X	X	X	Limited current availability Probably high future potential
					Monitoring/f'casting	X	X	X			Requires pesticides
					Irrigation mgmt	X	X				Region specific
					Biocontrol					X	
4	4	2	3	Early Blight (A. solani)	Sanitation	X	X				
					Resistance	X	X	X	X	X	Within market constraints
					Monitoring/f'casting	X	X	X			
					Irrigation mgmt	X	X	X			
					Fertility mgmt	X	X	X			Ground water concerns
					Biocontrol	X	X				
					Rotation	X	X			X	Economics? impact to other practices, limited land.
3	4	0	1	Potato Leafhopper	Biocontrol					X	Efficacy w/o pesticides?
					Forecasting				X	X	Need soaps, oils?
					Resistance				X	X	
4	4	4	0	Sprouting	Reduce temp.	X	X	X	X	X	Needs add'l cultivars
					Cultivars (long dormancy)	X	X	X	X	X	Market constraints
					Biocontrol					X	Currently quite limited

Regional Importance w/o Pesticides[a]

Impl.[b]

Research Timeframe[c]

Table 2 (continued).

Regional Importance w/o Pesticides[a]				Pest	Alternative	Impl.[b]		Research Timeframe[c]			Comment
NE	C	W	S			T	O	1-5	5-10	10+	
4	3	3	3	Vine Desiccation	Mowing	X	X				
					Flaming	X	X				
					Herbicidal soap	X			X		Registration? economics?
					Cultivar maturity					X	
					Biological				X	X	Region specific,tuber effects?
2	4	4	4	Verticillium	Sampling/forecast	X		X			Region specific
					Solarization	X		X			Sprayable films?
					Cover crops	X		X			Economics?
					Irrigation mgmt	X	X	X			
					Rotation	X	X	X			Limited land?
					Biocontrols				X	X	
					Resistance	X		X	X	X	Market constraints
4	4	4	0	Colorado Potato Beetle	Delay planting	X	X	X			Economical? Limited efficacy
					Rotation	X	X				Limited efficacy
					Early harvest	X	X				Economical efficacy?
					Flaming	X	X	X			
					Trap crops	X		X			
					Biocontrol				X	X	
					Vacuuming				X	X	Efficacy?
					Host resistance				X	X	Promising
					Straw mulch	X	X	X	X		Economics
					Microbials	X	X	X	X	X	Needs resistance mgmt

144

Table 3. Estimated increased difficulty expected from various potato pests of lesser importance in the absence of current pesticides.

Pest	NE	Regional Importance[a] C	W	S
Potato tuber moth	0[b]	0	2	1
Leafminers	0	0	0	2
Whiteflies	0	0	0	2
Thrips	0	0	0	2
Corky ringspot virus	0	0	1	3
Pythium	2	0	1	0
Psyllid	0	0	1	0
Mites	0	0	1	0
Grasshoppers	0	0	1	0
Bacterial wilt	0	0	0	4
Rhizoctonia	2	2	1	1
Wireworms	1	1	1	4
White grubs	1	1	0	0
Sclerotinia	1	1	2	3
Scab	2	1	0	0
Armyworms	0	0	1	1
Corn borers	1	2	0	0
Fusarium	2	1	2	0
Silver scurf	2	1	0	0

[a] Regional importance NE = Northeast; C = Central; W = West; S = South

[b] Scale of expected increased difficulty; 0 = no increase; 4 = very large increase

Implementation Needs

Many factors must be met to construct a successful IPM program. The following factors were considered important.

- The grower needs to be a central figure in the development of an IPM program. All persons (growers included) dislike regulations handed down from above. At the time of implementation, direct personal contact with growers should be maintained to facilitate communication and dialogue.

- There needs to be better marketing and integration of IPM components. Use of electronic technology in IPM can be greatly improved, and strategies for developing and marketing IPM programs need to be investigated and implemented.

- EPA should help maintain a wide range of options in all pest management tactics. A diversity of tactics (including a diversity of pesticides) is likely to enable more stable pest management. If a more restricted use of certain pesticides would maintain their availability, then implementing restrictions should be carefully considered.

- Processors and growers who support and practice IPM should be recognized and those who do not should be encouraged to do so.

- The labeling of vegetables produced according to IPM standards should be studied to determine the feasibility and probable impact on IPM adoption. IPM standards would give innovative growers a visible goal.

- Incentives should be provided to growers for adopting IPM practices. Successful scenarios need to be highlighted and successful adopters need to receive public recognition.

- Better support for IPM professionals/consultants should be provided in terms of workshops and seminars.

- There should be enhanced cooperation and collaboration among the USDA, EPA, and state agencies, who need to provide consistent, visible support to IPM. These groups need to share ownership in development and implementation of IPM programs. Existing regional IPM groups should be expanded for broader representation.

- USDA and EPA need to thoroughly understand grower practices before promulgating laws, regulations and label changes that seriously impact production and the use of IPM practices.

- Pesticide labels should emphasize the use of IPM methodology such as field and environmental monitoring, spraying only when necessary, use of thresholds, reduced rates, etc.

- There should be enhanced regionalization of IPM, so that states with limited resources can benefit from research results and implementation programs developed by other states in the region.

- IPM programs of the future will continue to substitute knowledge and management for pesticides. This approach will involve greater integration of many tactics with reduced reliance on any single "silver bullet" tactic.

Concluding Comments

Greater involvement by a larger segment of the United States population will significantly enhance IPM adoption. Ecologists, consumers and conservationists should be consulted early in the development of IPM programs. Producers must be central in the development of an IPM program, and their advice sought continually as the program develops. Consultants, extension personnel, industry personnel and researchers need to be involved in a dynamic dialogue during the research, development and implementation phases of an IPM program. Educational programs should be targeted at brokers, buyers, bankers, consumers, regulatory and funding agencies (EPA, USDA, FDA, state and local agencies). Specific educational programs should be developed for nonadopting growers. IPM will achieve greater success if the ownership of the concept and program is very broadly based.

Additional disciplines need to contribute in a basic way to the development of IPM programs. Sociological and economic issues need to be addressed at early stages of development. Issues such as farm worker safety, food safety and environmental conservation need to be priority goals. Systems science can contribute in an important way by enhancing the efficiency of IPM programs through the use of expert systems and programs which allow optimization.

The number of different-mode-of-action pesticides is quite limited for all groups of pests in potato agroecosystems. Retention of a diversity of pesticides is important because these provide the selection necessary for optimal use of all pest management tactics. With additional research and greater adoption of IPM programs, there will be significant efficiencies in the use of all pest management practices, including pesticides. However, decreases in the number of different types of pesticides will create significant problems with management of pesticide-resistant pests, and with certain cultural practices. For example, the elimination of the EBDC fungicides would severely constrain rotational schemes in many potato agroecosystems. The constraint comes from the fact that some crops in the rotational scheme for potatoes cannot be planted within 12 months on land which supported a crop sprayed with the replacement fungicide, chlorothalonil (12 month plantback interval). As pest management programs become more and more effective and necessary, it would be possible to move to something resembling prescriptive use of these tactics. In this event, there will need to be a greater number of well trained pest management practitioners.

VEGETABLE IPM
EXAMPLE: TOMATOES

As is the case for all vegetable crops, quality has always been of primary importance to tomato producers and handlers. In addition, pesticide residues have recently become a significant issue for consumers, growers, retailers and the tomato industry. Tomatoes are grown on almost 5.8 million acres worldwide, and on 700,000 acres in the United States.

Tomatoes share many similarities to other vegetable crops in terms of markets and cultivation. Tomatoes are produced both for processing and the fresh market. Consumption of both has risen dramatically in the past two decades with changes in diet. In the United States, processed tomatoes are largely produced in California, with minor production elsewhere especially in the East and Midwest. Fresh market tomatoes are produced for local markets in many areas of the United States, but significant production for shipment occurs in Florida and California in particular. Postproduction handling is important to insure a safe and high quality processed product, and to meet consumer demand for fresh tomatoes at times of the year when production is limited to only a few production areas. Although considered a minor crop by pesti-

148

cide registrants and USDA farm support programs, tomatoes represent a major vegetable crop.

Pest problems of tomatoes are many and vary with season, production region and cultivation practices. Almost 80 specific pest problems were identified and discussed by our workgroup. Late season tomatoes are exposed to more severe insect pest populations than those planted earlier, and traditionally receive more insecticide treatments. Disease pressure also increases in drier production areas such as California because of wetter conditions in the late summer and fall, although significant moisture at any time can result in damage. Areas where rainfall and high humidity occur during the production season typically require some disease control intervention to maintain a viable crop. Tomatoes can be seeded or transplanted. Seeded tomatoes are more affected by seedling insects, diseases and weeds, and are not suitable for many areas in which the crop must be produced in a shorter time period. Transplants in turn can have problems induced at the nurseries.

As is the case for all vegetables, tomatoes require intensive pest management to preserve yields and to meet damage standards imposed by retailers, processors, packers and governmental agencies that in large part are a response to consumer expectations and public health concerns. Processed tomatoes must meet state, federal and industry grading standards which indicate acceptable levels of insect and fungal damage. Processed tomatoes intended for whole pack instead of paste must be of even higher quality. Fresh market tomatoes are highly valued by consumers for their appearance, and those which are damaged by insects or disease are not acceptable at the retail level. As a result, pesticides have become an integral part of the production process. Some organic tomatoes are produced both for processing and the fresh market, but this acreage is quite limited and organically acceptable pesticides are still necessary to control pest problems.

IPM programs in various parts of the country have enhanced the management of many pest problems, and in some cases have reduced the amount of conventional pesticides applied for key pests by as much as 60% over the past decade. Additional reductions can be expected. However, it is essential that more effort be made to understand crop and pest ecology, and to develop IPM strategies for optimum pesticide use and nonchemical alternatives.

Implications of Removing Pesticides

In most cases, elimination of pesticides will create significant production problems in specific areas, and will increase the variability of production in others. It is likely that significant shifts of crop distribution will occur from such action, and that producers will be differentially impacted. Production in some areas of the United States might cease entirely except for local and seasonal markets. The cost of production in general will undoubtedly increase. It is quite possible that most of the United States market for tomatoes would be filled by foreign products which are produced inexpensively and with pesticides. This is underscored by the fact that the production of both processed and fresh market tomatoes is increasing relative to the United States in many parts of the world.

The implications for the elimination of all pesticides in tomato production on all major pests and groups of pests were evaluated regionally. Alternative tactics were identified for the most important pests, and the technical and operational feasibility of each was assessed (Table 4). If additional research is needed to develop the tactic or to enhance its effectiveness, the time expected to obtain implementable results was estimated (Table 4). Some of the alternative tactics are quite feasible (i.e., host plant resistance, microbial pesticides, etc.). Others are more speculative. Some tactics are immediately implementable, but would be more expensive (i.e., hand removal of weeds). The implications of removal of pesticides on some pests would be less severe (Table 5). This may be for three very different reasons: 1) the pest is of only occasional importance, 2) nonchemical controls are already the preferred method of control, or 3) no known pesticides control the particular pest. When a pest is only occasionally known to cause significant damage, it might be expected that differential impacts would result for producers. This would increase the variability (or uncertainty) of production both over time and geographic distribution.

Table 6a indicates the regional importance of specific weeds or groups of weeds. The composition of the weed complex in a given field is significant in choosing which herbicides should be applied, and impacts of rotations and other nonchemical tactics that might be applied. Solanaceous weeds, dodder and perennial weeds are of particular concern. Table 6b presents potential weed control tactics that could be used to address elements of the tomato weed complex, as well as their operational feasibility and research timeframe.

It is assumed for purposes of analyzing the potential of tomato pest control without conventional pesticides that some pesticides would continue to be available including oils, elemental and inorganic compounds (such as sulfur, copper and bicarbonates), microbial pesticides (including bacteria, viruses and fungal antagonists), and semiochemicals (such as pheromones).

150

It is obvious that tomatoes are attacked by many pests, some with potentially devastating implications if left uncontrolled. The importance of tomato pests differs considerably among regions and uses of the crop. IPM tactics applied must then be highly site specific. Pesticides reduce variability in the production system for many pests and in particular sites, making yield and quality more predictable. They are also convenient and inexpensive for producers to use relative to alternative methods, hence their widespread adoption at present. Many nonchemical IPM tactics have been identified that are not only technically feasible but currently implementable. However, most require significant research and development before they can be optimally used. In addition, many of the IPM tactics are cultural manipulations or biological controls, not products that would lend themselves to marketing by industry. Therefore, the role of extension educators and private consultants is crucial.

Table 4. Projected impact of elimination of all pesticides for selected pests of tomatoes in various regions of the United States and identification of potential alternatives with estimates of research effort needed to implement the alternatives.

| Regional Importance w/o Pesticides[a] | | | Pest | Alternative | Impl.[b] | | Research Timeframe[c] | | | Comment |
NE	SE	W			T	O	1-5	5-10	10+	
2	2	4	Verticillium	resistance	X	X	X	X	X	some Calif. cultivars resistant
				rotation	X	X	X			large acreage, expense
				sanitation	X	X	X			
				biocontrol					X	
2	4	0	Bacterial spot	transplant certification	X	X	X			mechanism needed
				resistance	X	X		X	X	some Calif. cultivars resistant
				rotation	X	X	X			large acreage, expense
				sanitation	X	X			X	
2	1	4	Blackmold	resistance	X	X	X	X	X	economics
				harvest date	X	X				economics
				earlier harvest	X	X				economics
				canopy management	X		X			

a Regions of the country are Northeast (NE); West (W); and Southeast (SE); 4 = large impact of eliminating pesticides; 0 = no impact of eliminating pesticides.

b Status of implementation T = technically possible; O = operationally possible

c Years to point of field implementation.

152

Table 4 (continued).

Regional Importance w/o Pesticides[a]			Pest	Alternative	Impl.[b]		Research Timeframe[c]			Comment
NE	SE	W			T	O	1-5	5-10	10+	
4	0	0	Anthracnose	rotation	X		X			wide host range; soil endemic
				resistance	X	X		X	X	
				monitoring	X	X	X			
4	3	2	Early blight	rotation	X	X	X			
				resistance	X	X	X	X	X	expense?
				fertility	X	X	X			fresh market now, processing
				monitoring	X	X	X			later
				irrigation management	X	X	X			environmental effects?
				biocontrol					X	
1	2	4	Late blight	resistance	X				X	
				sanitation	X	X				
				rotation	X	X				economics
0	4	4	Gemini viruses	biocontrol (vectors)	X		X			
				resistance	X		X	X	X	economics?
				rotation	X	X	X			
				cultural practices	X	X	X			
				host destruction	X	X	X			needs areawide implementation
				vine destruction/flaming	X	X	X			needs areawide implementation

153

Table 4 (continued).

Regional Importance w/o Pesticides[a] NE	SE	W	Pest	Alternative	Impl.[b] T	O	Research Timeframe[c] 1-5	5-10	10+	Comment
0	4	2	Nematodes	resistance	X	X	X	X	X	regional
				rotation	X	X			X	expense?
				biocontrol	X			X	X	
				cover crops	X			X		
				solarization	X	X				marginal efficacy
3	1	1	Phytophthora	raised beds	X	X		X		
				irrigation mgmt.	X	X		X		expense?
				land-leveling	X	X		X		expense?
				rotation	X	X		X		
				resistance	X			X	X	
1	4	0	Fusarium crown rot	rotation	X	X	X			expense?
				biocontrol	X		X	X	X	
				resistance	X		X	X	X	
				irrigation mgmt.	X	X	X			
3	1	3	Flea beetles	mechanical/stick'em	X	X	X			equipment needed
				scraping soil	X	X	X			cost, efficacy unknown
				resistance	X	X		X		
				cryolite	X	X		X	X	efficacy unknown

Table 4 (continued).

| Regional Importance w/o Pesticides[a] | | | | | Impl.[b] | | Research Timeframe[c] | | | |
NE	SE	W	Pest	Alternative	T	O	1-5	5-10	10+	Comment
2	4	1	Thrips	resistance	X				X	lack of neighbor control
				isolation	X	X	X			efficacy unknown
				microbial	X	X	X			
0	4	4	Sweetpotato whitefly	soaps	X	X	X			moderately effective
				predators/parasites	X	X	X	X	X	unknown efficacy
				microbial	X	X	X	X	X	dry climate may limit
				resistance	X		X	X		limited at present
				deterrents	X			X		registration required
				botanicals	X		X	X		registration; efficacy; safety
				trap crop	X					expense; efficacy
1	2	1	Cabbage looper	Bt	X	X				good control
1	1	1	Hornworms	Bt	X	X				good control
0	3	2	Leafminers	conservation of predators/parasites.	X	X				typically endemic
0	0	4	Russet mite	sulfur	X	X				good control
				resistance				X		
				weed control	X		X		X	

155

Table 4 (continued).

Regional Importance w/o Pesticides[a] = NE, SE, W · Impl.[b] = T, O · Research Timeframe[c] = 1-5, 5-10, 10+

NE	SE	W	Pest	Alternative	T	O	1-5	5-10	10+	Comment
3	0	4	Aphids (potato, green peach, cotton, citrus)	predators/parasites	X	X	X			moderately effective
				oils	X	X	X			moderately effective
				soaps	X	X	X			
				resistance	X	X	X	X		registration required
				alarm pheromones	X	X	X	X		registration required
				deterrents	X	X	X			expense/efficacy
				reflective mulch	X	X	X			
0	1	1	Tomato fruitworm	Bt	X	X	X		X	improvement expected/needed
				viruses	X	X	X			registered
				nematodes	X	X	X			unknown efficacy
				pheromones	X		X			immigration; registration
				parasites	X		X			improve quality
				resistance	X			X		includes Bt gene
2	3	4	Beet armyworm	resistance	X		X	X		includes Bt gene
				Bt	X	X	X	X	X	improvement needed
				viruses	X	X	X			registration required
0	4	4	Tomato pinworm	pheromones	X	X	X			not adequate alone
				host free period	X	X				expense

156

Table 4 (continued).

Regional Importance w/o Pesticides[a]			Pest	Alternative	Impl.[b]		Research Timeframe[c]			Comment
NE	SE	W			T	O	1-5	5-10	10+	
1	2	4	Stink bugs	parasites	X	X	X			improvement needed
				aggregation pheromone	X			X		
4	1	0	Colorado potato beetle	rotation	X	X	X			large acreage/expense
				flaming	X		X			plant damage
				trap crops	X	X	X			expense/efficacy
				biological control	X			X	X	
				resistance	X	X		X	X	
				microbials	X		X			resistance
				barriers	X		X			efficacy
				mulches	X		X			expense/efficacy
				cryolite bait	X	X				efficacy

Table 5. Estimated impact upon tomato production expected for various pests of lesser importance in the absence of current pesticides.

Pest	Regional Importance[a]		
	NE	SE	W
Buckeye rot	2	1	1
Southern bacteria wilt	0	3	0
Pith necrosis	1	2	0
Canker	1	2	1
Alternaria stem canker	0	1	0
Tomato spotted wilt virus	1	3	1
Cucumber mosaic	1	1	1
Poty viruses	0	2	0
Septoria leaf spot	2	0	0
Southern blight	0	2	0
Leaf mold -- Cladosporium	2	2	0
Gray mold	2	1	1
Target spot	0	3	0
Powdery mildew	0	0	4
Alfalfa mosaic	0	0	0
Curly top	0	0	3
Pseudo curly top	0	2	0
Tobacco mosaic	0	0	0
Pythium	1	3	1
Rhizoctonia	1	2	1
Corky root	0	2	2
Fusarium wilt	0	2	1
Verticillium	0	1	2
Sclerotinia	1	2	0
Cutworms	1	1	1
Seedcorn maggot	0	0	1
Wireworms	0	3	1
Symphylan	1	0	2
Greenhouse whitefly	1	1	1
Spider mite	1	2	0
Tobacco budworm	0	1	1
Western yellowstriped armyworm	0	1	3
Potato tuberworm	0	0	1
Mole crickets	0	3	0
Southern armyworm	0	2	0
Fall armyworm	1	1	0
The armyworm	2	0	0
European corn borer	1	1	0
Banded/Cloudy-winged whitefly	0	2	0

[a] Regions of the country are Northeast (NE); West (W); and Southeast (SE); 4 = large impact of eliminating pesticides; 0 = no impact of eliminating pesticides

Table 6a. Projected impact on tomatoes of elimination of all herbicides for selected weeds in various regions of the United States and identification of potential alternatives with estimates of research effort needed to implement the alternatives.

Regional Importance w/o Pesticides[a]			Pest	Alternative from Weed Control List (Table 6b)
NE	SE	W		
1	0	3	Winter broadleaves	2-22
4	4	4	Annual grasses	1-22
4	0	3	Velvetleaf	2-22
3	3	0	Ragweed	2-22
4	0	0	Galinsoga	2-22
0	4	0	Parthenium	2-22
3	3	4	Purslane	2-22
3	1	3	Annual morningglory	2-22
3	4	4	Solanaceous weeds	2-22
2	3	4	Yellow nutsedge	2-5, 7-9, 12-16, 18-22
0	0	0	Purple nutsedge	2-5, 7-9, 12-16, 18-22
2	0	4	Field bindweed	2-5, 7, 12, 13, 16, 18-22
1	3	4	Johnsongrass	1-7, 12, 16
0	3	1	Bermudagrass	18-22
0	1	2	Dodder	2-7, 15-16, 18-22
3	4	4	Pigweed	2-22
0	4	1	Eclipta	2-22
0	3	0	Kudzu	2-5, 7, 12, 14, 17-18, 22
0	3	0	Smooth vetch	2-22
0	4	0	Sesbania	2-22
1	3	1	White & yellow clover	2-22

[a] Regions of the country are Northeast (NE); West (W); and Southeast (SE); 4 = large impact of eliminating pesticides; 0 = no impact of eliminating pesticides

Table 6b. Projected impact on tomatoes of elimination of all herbicides for selected weeds in various regions of the United States and identification of potential alternatives with estimates of research effort needed to implement the alternatives.

Alternative	Impl.[a] T	Impl.[a] O	Timeframe 1-5	Timeframe 5-10	Timeframe 10+	Comments
1. Weeder geese	X	X	X			Water, dogs
2. Rust	X				X	
3. Insects	X				X	
4. Nematodes	X				X	
5. Mycoherbicides	X				X	Lack broad spectrum
6. Sanitation during harvest	X		X			Cost
7. Rotation	X	X	X			Economics?
8. Living mulches	X			X		High inputs, pest increases?
9. Plant density/arrangement	X			X		Equipment, difficult cultivation
10. Mulches--organic	X	X	X			Pest increases, cooler soils, irrigation problems
11. Plastic mulches	X	X				Cost, disposal
12. Cultivation	X	X	X			Within row, fuel costs, rain, compaction
Cultivation--night	X		X			Within row, fuel costs, rain, compaction
Cultivation--robotic vision	X		X			Within row, fuel costs, rain, compaction
13. Transplants	X	X				Cost, pest transmission
14. Solarization	X	X				Cost, timing/economics, disposal, rain
15. Cultivar selection	X			X		
16. Subsurface drip	X	X	X			Rotation limited, cost, disposal
17. Hand hoing	X	X				Cost, labor availability
18. Flaming	X			X		Fuel cost
19. Planting date	X			X		Cultivar, weather, contracts
20. Nutrient modification	X				X	Biology poorly known
21. Monitoring/prediction	X			X		Difficulty monitoring seeds/cost, heterogeneity
22. Solar flaming	X			X		Sunny conditions

[a] Status of implementation T = technically possible; O = operationally possible

Implementation Needs

Successful development and use of an IPM system for tomatoes would be enhanced if the following factors were considered:

- Fresh market and processing tomatoes require different production practices, and should not be treated the same for evaluation of pesticide or pest management impact and use.

- EPA and state regulators can seriously affect the ability to use crop rotations or resistance management strategies by adopting various restrictions. They need to thoroughly understand production practices before implementing laws, regulations or label changes.

- More stringent restrictions on certain pesticides should be considered if they would address safety concerns yet maintain availability for specific purposes.

- Pesticide labels should be standardized so that they are in the same format or perhaps color-coded by type of instruction.

- Non-English labels should be required at least for safety information as many tomato producers and other individuals who work with labor intensive vegetable crops do not speak English as their primary language.

- Pesticide labels should emphasize the use of IPM practices such as pest and environmental monitoring, effects on beneficial organisms, use of thresholds, potential for using reduced rates, etc.

- Producers, regulators, receivers, brokers and consumer groups must reach a consensus on what defect levels are appropriate for a particular use of tomatoes. However, this issue may prove unresolvable because of public dislike for defects and the resulting conflict between producers and receivers or brokers.

- Enhanced cooperation should occur between USDA, EPA and state agencies, and these organizations should provide consistent, visible support for IPM development and implementation. Crops not supported by USDA funds such as tomatoes should receive greater consideration in research and implementation funding.

- Producers need to be part of the process of developing and implementing an IPM program.

- Producers need to talk to each other as well as educators and regulators. Areawide planning would be useful to manage many pests of vegetable crops such as tomatoes where production seasons can overlap resulting in increased pest populations.

- Integrating IPM components and more efficient delivery is essential. Electronic technologies hold great promise for this purpose and should be encouraged.

- Tomato processors, brokers, growers and retailers who support and practice IPM should be recognized by public agencies, consumer and environmental groups, and those who don't should be encouraged to do so.

- Incentives, from both public agencies and the private sector, should be provided to growers for adopting IPM practices.

- Promote the plant health consultant profession, and train growers to use the recommendations provided.

- Better educational support for IPM consultants should be provided.

- IPM adoption or adaptation by producers should be measured, and rigorous and acceptable methodology for such studies developed.

Concluding Comments

The tomato plant is the integrator of all the external factors that it is exposed to from the environment, production practices and the pest complex that attacks it. Similarly, the tomato producer is the integrator of finances, location, contractual obligations to bankers, receivers or brokers, regulations and production information received. Because IPM is a complex set of tactics which must be organized into a system that addresses the needs of the individual tomato producer and society as a whole, it must be recognized that the producers must address many concerns other than their pest problems and that society has concerns other than those of the producer. IPM educational programs should be targeted at businesses and agencies that influence a producer's pest management practices, not just to producers and their consultants. In addition, the public must learn to understand the difficulties of growing food crops such as tomatoes. The issues of pesticide impacts upon farm workers, the environment and the individual farmer must be adequately addressed at early stages of program development. The successful development and implementation of IPM tactics can help to mitigate many of these issues which relate to pesticide use.

VEGETABLE CROPS WORKING GROUP

The following individuals participated in the development of information presented in this report through our survey or in meetings to develop the crop examples. Many of them also reviewed the manuscript which is a summary of their thoughts.

George Abawi
NYS Agricultural
Experiment Station

Paul A. Backman
Auburn University

Ed Beckman
California Tomato
Board
Fresno, CA

Richard L. Clark
ROCAP
Guatamala City

John Edelson
Texas A&M
University

Peter B. Goodell
University of
California
Cooperative
Extension

Juan Ancisco
Edinburgh, TX

Steven S. Balling
Del Monte Foods
Walnut Creek, CA

Dan Botts
Florida Fruit &
Vegetable Assoc.
Orlando, FL

H. Dillard
NYS Agricultural
Experiment Station

David N. Ferro
University of
Massachussetts

E. J. Grafius
Michigan State
University

Sandra Archibald
Stanford University

Jerry A. Bartz
University of Florida

James Breinling
Gerber Products
Company
Fremont, MI

Jere D. Downing
Ocean Spray
Cranberries, Inc.
Middleboro, MA

Deborah Fravell
USDA-ARS
Beltsville, MD

Gary Harman
NYS Agricultural
Experiment Station

Frederick A. Hegele
General Mills, Inc.
Minneapolis, MN

C. M. Hoy
Ohio State
University

P.A. Koepsell
Oregon State
University

Thomas Lanini
University of
California, Davis

Ron Mau
University of Hawaii

Madeline Mellinger,
President
National Alliance
Independent Crop
Consultants
Jupiter, FL

Albert O. Paulus
University of
California
Riverside

Mary L. Powelson
Oregon State
University

Warren Springer
Northrup King
Minneapolis, MN

Paul C. Vincelli
University of
Kentucky

Richard C. Henne
Campbell Research
Napoleon, OH

Marshall W. Johnson
University of Hawaii

Steven T. Koike
University of
California
Cooperative
Extension

Michael Lewis
Ore Ida Foods
Ontario, OR

Gary McKinsey
Meyers Tomatoes
King City, CA

Gene Miyao
University of
California
Cooperative
Extension

Curt H. Petzoldt
NYS Agricultural
Experiment Station

Charles Rivara
California Tomato
Research Institute
Fresno, CA

W.R. Stevenson
University of
Wisconsin

J.F. Walgenbach
North Carolina State
University

Michael Hoffmann
Cornell University

George Klacan
Pillsbury Inc.
LaSueur, MN

Frank F. Laemmlin
University of
California
Cooperative
Extension

Pam Marrone
Entotech, Inc.
Davis, CA

Charles Mellinger
Glades Crop Care
Jupiter, FL

Joseph Panetta
Mycogen
Corporation
San Diego, CA

Kenneth L.
Pohronezny
University of Florida

Douglas I. Rouse
University of
Wisconsin

J.T. Trumble
University of
California
Riverside

Craig V. Weakley
San Tomo Group
Sacramento, CA

R.D. William
Oregon State
University

P.H. Williams
University of
Wisconsin

G.W. Zehnder
Virginia Polytechnic
Institute

T.A. Zitter
Cornell University

Chapter 6

CONSTRAINTS TO THE IMPLEMENTATION AND ADOPTION OF IPM

Edward H. Glass
Department of Entomology
New York State Experiment Station
Cornell University
Geneva, NY 14456

Integrated Pest Management (IPM) is difficult to define, not because it is so complex or abstract, but because it is an approach to pest control. It is a strategy rather than a specific and exact methodology. Its strength is in its adaptability in one form or another to all pest problems. Definitions often get bogged down in adding specifics which apply in some situations and not in others. One of the pioneers in developing IPM and one who did much of his research in the field defined IPM as follows: "IPM is the balanced use of such measures, cultural, biological and chemical, as are most appropriate in the light of careful study of all factors involved" (21) . It is simple, yet encompasses the breadth of the concept without getting involved with specifics which may or may not be applicable in a given situation.

As has been pointed out in one of the commodity team reports, IPM systems are site specific. IPM systems do and must vary by crop, by cropping system and even from one part of a farm to another. They can encompass relatively small ecological areas up to entire regions. Further, IPM systems must be dynamic rather than static because of the evolutionary adaptation of pests to adverse environments such as pesticides, resistant host plants, and cultural and biological controls. There should be continuous monitoring and modifications as needed. The latter most often are supplied through additional research and demonstrations. In developing IPM systems generally there is a continuum of activities and efforts from basic through applied research to field development and finally implementation by farmers or other users. This process is affected by many constraints which limit the development and implementation of IPM.

The list below is a compilation of constraints identified by the four committees dealing with specific commodities (corn/soybeans, cotton, fruit, vegetables). The constraints are categorized in one or more of the following areas:

Research - (technical)
Extension - (communications)
Education - (public, farmers, general)
Training - (IPM specialists)
Institutional - (universities)
Regulation
Policy
Economic

Discrete separations do not exist among some of these categories. For example, some IPM research is done by extension specialists and some extension by field oriented researchers. Some scientists have joint extension/research appointments. The distinction seems warranted, however, because the programs are administratively divided at both the state and federal levels. Likewise, there is a blurred distinction between regulation and policy. In this case, the thrust of regulatory constraints involves federal and state agencies whereas policy constraints lie predominantly in administrative and legislative bodies.

The team(s) identifying each constraint are indicated by V for vegetables, C for cotton, C/S for corn/soybean and F for tree fruit following each constraint. The constraints are abbreviated in most cases from the fuller coverage in the reports. No attempt has been made to prioritize them.

CONSTRAINTS

Research (Technical)

- Lack of understanding of crop and pest biology and interactions between and among them. V, C/S, F

- Lack of field-oriented IPM research on the biology of pests and multiple pest interactions. C/S, F

168

- Need to simplify IPM methodologies, especially monitoring and sampling pests and beneficials. V, C/S

- Lack of systems level applied research. C/S

- Lack of economic data. This need should receive the same attention and consideration as statistical accuracy. V, C/S

- Inconsistent results with biological control agents. Lack of information on the biology of such agents, environmental tolerances, application requirements, storage and mass production. C, F

- Lack of factual data on pest levels and damage needed to prioritize IPM research and development efforts. C/S

- The large plot requirements and high costs of field testing biologically oriented IPM tactics. C/S, F

- Underground pests (nematodes, grubs, pathogens) are difficult to research. C/S

- The evolution of pest resistance for most control tactics requires continuing research and demonstrations of new strategies and tactics. C/S, F

- Biologically oriented IPM requires 5 to 10 years of labor intensive research efforts, especially rotations and new approaches such as semiochemicals. C/S

- Need for more interdisciplinary research. C/S

- Lack of funding and shortage of personnel to conduct the needed site-specific research and demonstrations. V, C, C/S, F

Extension/Demonstration

- Producer perception that IPM is riskier and more expensive than conventional controls. Preference for immediate solutions. Low value crops do not favor long-term solutions. Traditional emphasis on control rather than management. V, C, C/S, F

- Lack of communication and promotional channels for biologically oriented IPM comparable to those for pesticides. V

- Shortage of trained IPM personnel to develop and transfer IPM technology at the local level. C, C/S, F

- Pro-pesticide attitude of some extension agents and consultants. C

- Lack of extension demonstrations and education programs due to insufficient support. C

- Lack of public acceptance for. C/S

- Lack of awareness of pest resistance problems and the need for solutions plus the lack of trained experts at all levels on resistance management. C/S

- Inability of Cooperative Extension Service to demonstrate tangible increased producer profits through the use of biologically oriented IPM. C/S

Education (Public, Producers, Consumers)

- Public alarm about genetically engineered organisms which results in overly restrictive regulations (also for all pesticides). C/S

- Public demand for nonblemished foods. V, F

- Perception that IPM is riskier and more expensive than conventional controls (also listed under Extension). C/S, V

- Societal concerns related to all pesticides. C/S

- Decreased understanding and appreciation of agricultural research need and benefits by the public. C/S

Training (IPM Specialists)

- Need for universities to strengthen IPM training programs and develop curricula required to train people to work at regional as well as local levels. V, C, C/S

- Shortage of scientists dedicated and trained to do interdisciplinary research and extension on biologically oriented IPM. C/S

- Lack of support for breeder education programs. C/S

- Need for more well-trained private sector IPM personnel and training in IPM monitoring and decision making. V, C/S, F

Institutional (University)

- Lack of support for interdisciplinary collaboration in IPM research, extension and teaching (given low priority). V, C, C/S

- Degree programs structured toward narrow expertise rather than broad knowledge of cropping systems. V, C, C/S, F

- Administration and funding agencies favor short-term (1 to 3 year) projects whereas biosystems research requires 5 to 10 years. C/S

- University tenure and promotion policies discourage long-term research required for applied ecological studies. C/S, F

- Lack of university/Environmental Protection Agency (EPA)/United States Department of Agriculture (USDA) funded regional IPM centers. V, C

- Cultural (rotation, etc.) research is management oriented and is given lower priority than high technology research. C/S

- Institutional emphasis is on basic research rather than on applied ecology. C, C/S

Regulation

- Pesticide plant-back restrictions (lack of a registration and residue tolerance for a rotation crop for a pesticide used on an earlier crop) limit crop rotation options. V

- Lack of adequate and appropriate regulation procedures. Lack of generic registration for microbials and semiochemicals across broad crop categories. V

- Interstate regulations of transport across state lines are constraints to biological control programs. V, C/S, F

- The time-consuming, expensive and unpredictable EPA pesticide review and registration process. C, C/S

- Lack of clear EPA guidelines for genetically engineered products. C/S, F

- Lack of state policies and regulations for area-wide IPM activities (e.g., stalk destruction of cotton). C

- Diversity and uncertainty of international regulations. C/S

- High cost of developing pest control agents due to regulations, especially for small producers and researchers. C/S, F

- Duplication of federal and state regulations of genetically engineered organisms. Lack of coordination. C/S, F

- Poor interpretation of efficiency standards for nontoxic controls. F

- Unreasonable restrictions on testing and plot sizes. F

- Unreasonable crop destruction requirements. F

- Regulations on the use of patented clones as parents in breeding programs. F

- Separate registrations package requirement for each strain of organism. F

- High nonspecific standards for insect parts in foods. V, F

- Lack of distinction between toxic and nontoxic materials in reviews conducted by same reviewer. F

- Lack of accelerated review process for biointensive control products (EUP's) F

Policy

- Lack of consistent policies in government agencies following biointensive IPM. Some discourage IPM. V, F

- Emphasis and support for program crops has diverted research, extension, and grant support away from vegetables and other nonprogram crops. V

- Lack of crop insurance. V

- Subsidy of monocropping and maximum yield incentives favor pesticide use. V, C

- High standards of FDA, USDA, state and food processing industry constrain biointensive IPM. V, F

- Lack of incentives for producers to try or to adopt biointensive IPM. C

- High capital demands for land use. C

- Emphasis on basic research with inadequate support of applied ecology by universities and federal funding entities. C

- Lack of policies and support/subsidies for IPM in sensitive areas (e.g., endangered species habitats). C

- Lack of coordination of agency policies which impinge on IPM, biological controls, etc. (Soil Conservation Service, water quality, endangered species, etc.) C

- Lack of patentability for biological control agents (except bioengineered). C/S

- USDA Competitive Grants, Low Input Sustainable Agriculture (LISA), etc. do not address needs for developing biointensive IPM. C/S

- Farm programs encourage farmers to make short term decisions rather than long-term ones which encompass the benefits of biointensive IPM. C/S

173

- Lack of long-term funding commitments for research, extension, and implementation activities. V, C/S

- Insufficient emphasis on preserving genetic diversity essential for breeding pest resistant cultivars. C/S

- Lack of coordinated effort to address pest resistance to all control tactics. C/S

- Lack of access to key personnel in regulatory agencies. F

Economic

- Economics of managing pests may favor chemical control (see tree fruit report). F

- Lack of incentives for manufacturers and farmers for biointensive IPM. F

- Many effective, inexpensive pesticides plus service available to producers. C/S

- Expense and length of time required to obtain registrations of new biointensive controls. C/S

- Lack of profitability for private sector research and development of biointensive products and technology. C/S

- Biointensive systems cannot be "packaged" for sale to farmers. Must be tailored to local conditions. C/S

APPENDIX 1

A survey of cotton industry representatives was made to help identify research and extension needs for development of biologically intensive IPM. Below is a fascimile of the survey.

December 4, 1990

MEMORANDUM

TO: Colleagues in the Cotton Industry

FROM: Ray Frisbie   Dick Hardee
      Department of Entomology   USDA, ARS, SIML
      Texas A&M University   P. O. Box 346
      College Station, Texas 77843   Stoneville,
                                      Mississippi 38776

SUBJECT: EPA/USDA Plan for a Public/Private Forum on
         Biologically Intensive Integrated Pest Management (IPM)

Never before has the future of cotton IPM been more challenging. Environmental concerns, economic concerns on the part of the pesticide industry, as well as pesticide resistance will critically affect the use and availability of pesticides on cotton and other crops in the future. The recent challenge of the pyrethroids by EPA is an all too sobering example of how quickly an entire class of insecticides can be placed in jeopardy.

A serious look into the future to identify pest management alternatives and techniques that will indeed allow us to produce cotton economically with fewer pesticides is the charge of the EPA/USDA Forum on Biologically Intensive IPM. *Biologically intensive IPM for this discussion relies primarily on biological control, host resistance, cultural management and the judicious use of environmentally safe pesticides.* As Co-chair persons of the Cotton Action Team, we are seeking a broad range of input. You and others are being contacted as leaders in cotton research and extension to help us describe what pest management in cotton production will look like in the next 10-15 years. The primary assumption is that we will likely have far fewer insecticides, fungicides, nematicides and herbicides with which to do our jobs. It is up to us to describe what research and extension needs will be required in a pesticide limited cotton production system.

In order to gain your valued and creative input, we have enclosed a brief questionnaire. As you address the questions, a good starting point might be to assume that we will have few if any pesticides in cotton production within the next 10-15 years. Please take care in addressing question number 4 which asks you to identify research needs in a pesticide limited environment. Try not to say it can't be done and use your imagination to arrive at research needs that may lead to biologically intensive IPM. For example, it is difficult to think of cotton production without or with very few herbicides. The sky is the limit. Go for it.

Your input will be included in a forum report. This report will be taken into consideration by EPA and USDA and will very likely determine future policy for funding of research and extension programs. Please feel free to consult your colleagues as you consider the questionnaire. This is a massive challenge for which no one person has the answer, but working together we can create solutions and opportunities. We would appreciate having your response by January 15th, 1991. We hate to rush you on such an important request, but we definitely need your input. We sincerely appreciate your help.

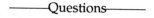
————Questions————

1. What biologically intensive IPM tactics (biological control, host resistance, cultural management) are currently available or have potential for immediate use?

2. What are the major constraints that prohibit the development or use of the above biologically intensive IPM tactics?

3. What research is currently underway that focuses primarily on biologically intensive IPM techniques?

4. What research is needed in the future to bring about biologically intensive IPM?

5. Specifically, what ways would you suggest to resolve pest resistance resulting from the pesticide use or anticipated resistance to genetically altered, transgenic plants?

6. What is the applied research role of the Cooperative Extension Service in the development and transfer of biologically intensive IPM?

7. What current state/federal policies constrain the use of biologically intensive IPM?

8. What changes would you suggest for question #7 above?

9. What state/federal regulations constrain the development or use of biologically intensive IPM?

10. What changes would you suggest for question #9?

11. In your opinion, what is the number one challenge facing cotton IPM in the next century?

LITERATURE CITED

1.   Anonymous. 1975. Pest Control: An Assessment of Present and Alternative Technologies. Volume II. Corn/Soybean Pest Control, National Academy of Sciences, Washington, D.C. 169 pp.

2.   Anonymous. 1990. Crop Values. 1989 Summary USDA National Agricultural Statistics Service, Agricultural Statistics Board, Washington, D.C.

3.   Council on Environmental Quality. 1972. Integrated Pest Management. Council on Environmental Quality, Washington D.C.

4.   Frisbie, R.E., and Adkisson, P.L. 1985. IPM: Definitions and current status in U.S. agriculture. Pages 234-262 in: Biological Control in Agricultural IPM Systems. M.A. Hoy and D.C. Herzog, eds. Academic Press, New York. 589 pp.

5.   Frisbie, R.E., El-Zik, K.M., and Wilson, L.T. 1989. The future of cotton IPM. Pages 413-428 in: Integrated Pest Management Systems and Cotton Production. John Wiley & Sons, New York. 437 pp.

6.   Georghiou, G.P. 1986. The magnitude of the resistance problem. Pages 14-44 in: Pesticide Resistance: Strategies and Tactics for Management. National Academy Press, Washington D.C.

7.   Good, J.M. 1973. Pilot Programs For Integrated Pest Management Conf., Kiev, USSR. Extension Service, Washington, D.C. ANR-5-15 (10-73).

8.   Gunsalus, J.L. 1990. Mechanical and cultural weed control in corn and soybeans. Am. J. Alternative Agr. 5(3): 114-119.

9.   Handy, R.B. 1896. History and general statistics of cotton. Pages 17-66 in: The Cotton Plant. U.S.D.A. Bulletin, Washington, D.C.

10.   Herzog, G.A., and Hardee, D.D. 1991. Highlights of the 1991 Cotton Insect Research and Control Conference. Pages 119-121 in: Proceedings of the Beltwide Cotton Conferences. D.J. Herber and D.A. Richter, eds. National Cotton Council of America, Memphis, Tennessee. 1031 pp.

11. Huffaker, C.B., ed. 1971. Biological Control. Plenum Press, New York. 511 pp.

12. Klonsky, K., Zalom, F.G. and Barnett, W.W. 1990. Evaluation of California's almond IPM program. Calif. Agric. 44(5): 21-24.

13. Michelbacher, A.E., and Bacon, O.G. 1952. Walnut insect and spider mite control in northern California. J. Econ. Entomol. 45:1020-1027.

14. National Cotton Council. 1990. Beltwide Cotton Production Res. Conf. National Cotton Council of America, Memphis, Tennessee. 1031 pp.

15. Norgaard, R.B. 1975. Evaluation of the pest management program for cotton in California and Arizona. Pages C1-59 in: Evaluations of Pest Management Programs for Cotton, Peanuts and Tobacco in the United States. R. Von Rumker, G.A. Carlson, R.D. Lacewell, R.B. Norgaard and D.W. Parvin, eds. Washington D.C. Final Report Contract No. EQ4ACQ36, Council on Environmental Quality. 589 pp.

16. Office of Technological Assessment (OTA), U.S. Congress. 1986. Technology, Public Policy, and the Changing Structure of American Agriculture. OTA-F-285. Washington D.C. 374 pp.

17. Rajotte, R.J., Kazmierczak, R.F., Lambur, M.T., Norton, G.W. and Allen, W.A. 1987. The National Evaluation of Extension's Integrated Pest Management (IPM) Programs. Publication 491-010 through 491-024. Virginia Cooperative Extension Service, Blacksburg, VA. 123 pp.

18. Stimmann, M.W., and Ferguson, M.P. 1990. Potential pesticide use cancellations in California. Calif. Agric. 44(4):12-16.

19. United States Department of Agriculture. 1988. Agricultural Resources-Inputs-Outlook and Situation Report. AR-9. Economic Research Service, Washington D.C. pp.

20. United States Environmental Protection Agency. 1986. Pesticide Industry Sales and Usage: 1985 Market Estimates. Washington D.C.

21. Way, M.J. 1977. Integrated control - practical realities. Outlook Agriculture 9:127-135.

22. Willey, W.R.Z. 1977. Regulatory needs for the implementation of IPM. Pages 60-66 in: Conf. Proc. New Frontiers in Pest Management. Sacramento, CA.